AGRICULTURAL SYSTEMS: ECONOMICS, TECHNOLOGY AND DIVERSITY

AGRICULTURAL SYSTEMS: ECONOMICS, TECHNOLOGY AND DIVERSITY

OLIVER W. CASTALONGE
EDITOR

Nova Science Publishers, Inc.
New York

For permission to use material from this book please contact us:
Telephone 631-231-7269; Fax 631-231-8175
Web Site: http://www.novapublishers.com

LIBRARY OF CONGRESS CATALOGING-IN-PUBLICATION DATA

Agricultural systems : economics, technology and diversity / Oliver W. Castalonge [editor].
p. cm.
Includes bibliographical references and index.
ISBN 978-1-60692-025-1 (hardcover : alk. paper)
1. Agricultural systems--Case studies. I. Castalonge, Oliver W.
S494.5.S95A374 2009
630--dc22

2008036569

Published by Nova Science Publishers, Inc. ✢ New York

CONTENTS

PREFACE

Agricultural systems include the development and application of systems methodology, including system modelling, simulation and optimisation; ecoregional analysis of agriculture and land use; studies on natural resource issues related to agriculture; impact and scenario analyses related to topics such as GMOs, multifunctional land use and global change; the development and application of decision and discussion support systems; approaches to analysing and improving farming systems; technology transfer in tropical and temperate agriculture; and the relationship between agricultural development issues and policy. This new book presents the latest research in the field.

Chapter 1 - This chapter is a synthesis of five empirical studies (Hansson, 2007a; Hansson, 2007b; Hansson, 2007c; Hansson, 2008; Hansson and Öhlmér, 2008) which are all parts of a research project explaining reasons for inefficiency in dairy production. The analysis was conducted in light of four themes: the overall structure of inefficiency, the farm itself, differences in operational managerial practices and farmer managerial capacity.

The research project was based on a dataset of 507 dairy farms in Sweden. Farm efficiency scores, based on accounting data of each farm, were calculated with the data envelopment analysis. Factors hypothesized to explain differences between efficient and inefficient farms were collected with a questionnaire which was answered by about 320 farmers. Data were also obtained from a dairy cow recording scheme conducted by the Swedish Dairy Association. The factors hypothesized to explain reasons for inefficient production were analyzed with regression models.

The results showed that the links between efficiency and commonly used critical success factors are typically weak, implying that the critical success factors cannot predict efficient farming. The results also showed that farm size affect different efficiency measures in opposite ways, implying that the overall effect is unclear. Regarding strategic factors, considered in light of the third theme (the farm itself), the results showed that more diversified farms are more efficient, and also that larger fields and bunker or tower silos lead to more efficient farms. Farms situated in a social context, where the farmer can discuss milk production with someone on a regular basis, were more efficient. The results further showed that factors adjustable in the daily work typically do not affect farm efficiency to any larger extent. Finally, the results showed that farmers who are educated in agriculture and continually up-date their skills, e.g., through study circles, generally run more efficient farms. The most fruitful insights into why some farms are inefficient were found in the themes the farm itself and farmer managerial capacity. The chapter ends with two suggestions for future research.

Chapter 2 - This chapter tests whether the Law of Proportionate Effects (Gibrat, 1931), which states that farms grow at a rate that is independent of their size, holds for the dairy farms in Northern Ireland. Previous studies have tended to concentrate on testing whether the law holds for all farms. The methodology used in this study permits investigation of whether the law holds for some farms or all farms according to their size. The approach used avoids the subjective splitting of samples, which tends to bias results. Additionally we control for the possible sample selection bias. The findings show that the Gibrat law does hold except in the case of small farms. This is in accordance with previous findings that Gibrat's law tends to hold when only larger farms are considered, but tends to fail when smaller farms are included in the analysis. Implications and further extensions, as well as some alternatives to the proposed methodology are discussed.

Chapter 3 - The economic loss of ammonia (NH_3) volatilization from chemical N fertilizers applied to farmlands worldwide is 11.6 billion US dollars per year. The economic impact of negative environmental effects resulted from NH_3 volatilization, i.e., formation of potent greenhouse gas (N_2O) and $PM_{2.5}$, is difficult to estimate but enormous. The Shannon's Information Theory was applied to the data collecteddd from our previous study using either ammonium sulfate [$(NH_4)_2SO_4$], ammonium nitrate (NH_4NO_3), potassium nitrate (KNO_3), or urea applied to four agricultural soils, i.e., Biscayne Marl Soil (BMS) and Krome Gravelly Loam (KGL) from Florida and Quincy Fine Sand (QFS) and Warden Silt Loam (WSL) from Washington, at two soil water regimes (20% and 80% field capacity, FC) and incubated at either 11, 20, or 29°C. Shannon rate (R_s g N ha^{-1} bit^{-1}) was defined as N-loss per unit area per Shannon entropy via NH_3 volatilization from the fertilized soils. The results showed that the R_s values across the four soils were 3-fold greater at 20% FC than that at 80% FC soil water regime. The BMS and KGL soils depicted similar R_s values, i.e., 2362.5 and 2378.9, while those for the QFS and WSL soils were only 1079.0 and 851.1 g N ha^{-1} bit^{-1}, respectively. The R_s values were 4178.6, 2863.3, 1502.3, and 21.7 g N ha^{-1} bit^{-1} for urea, $(NH_4)_2SO_4$, NH_4NO_3 and KNO_3, respectively. The environmental friendliness of either the tested soils, fertilizers, or soil water regimes, etc. were based on the R_s values, discussed.

Chapter 4 - The aim of this chapter is to evaluate the economic whole farm impact of the implementation of precision-farming technologies. Therefore, two farms were analyzed, which had implemented different precision agriculture technologies several years ago. The farms can be regarded as early adopters of precision agriculture technologies in two very different agricultural production areas. One of the two analyzed farms (1560 ha) is located in East Germany (Saxony-Anhalt), in a region with low annual rainfall and high fertile soils. In this farm nitrogen and potassium fertilizer applications, fungicides applications, and seeding are managed site-specifically following a mapping-approach, based on yield maps and aerial photographs. The other farm (150 ha) is located in Bavaria, in a region with high annual rainfall and comparatively fertile soils. In this farm the Yara-N-Sensor® is used for site-specific nitrogen management. The economic analysis is based on economic efficiency calculations for the implemented technologies, e.g., break-even-analysis and profit estimations. Furthermore, the two farmers were interviewed about the impact on their farm management and organization.

The implementation of the technologies provided economic gains for both of the farms. While the implementation of the sensor-based technology had only moderate effects on the overall farm management, the implementation of the mapping based system resulted in a far-reaching reorganization of the farm.

The authors conclude that a sensor-approach is easier to implement than a mapping approach. The implementation of a mapping-based site-specific management approach requires more skills, e.g., for analyzing site-specific data or creating application maps. Positive returns on the investments in the fertilizer technology can only be achieved when the technologies are used on sufficient acreage. This is easier to realize on big farms in eastern Germany or in terms of a collaborative use of the technologies than on comparatively small farms in southern Germany. However, in both regions the authors didn't find evidence that the innovation is adopted by neighboring farmers, which indicates that there are still considerable adoption constraints that limit the diffusion of precision farming technologies in Germany.

Chapter 5 - Traditionally Kazakhstan is a country of a transhumant farming system. After the independence from the Soviet Union, state farms were privatized. A large number of newly developed corporate farms were bankrupted through the privatization process, and household farms, a traditional livestock production system using common lands, emerged as private farms. On the other hand, since the prices of general commodities have been raised through the rapid economic growth supported by the oil boom, exporting livestock products becomes more difficult. In addition, it is likely that off-farm work opportunities provided by oil industries have made the traditional farming systems less important in rural economy.

The objective of the study is to quantify the role of the livestock sector in rural society and to elaborate public support strategies in Mangistau Oblast in Kazakhstan. The history of farm restructuring of Kazakhstan and the current trends of livestock sector in Mangistau Oblast were reviewed. Then, the current conditions of restructured farms were examined by household interviews (57 household and 43 peasant farms) and interviews with the leaders of village and corporate farms with regards to land use, production, revenue, marketing, and farmers' attitude.

The results showed that despite unfavorable climate conditions for agricultural production and long distance to market, the rural population in Mangistau Oblast is strongly attached to their lands and traditional livestock farming. Nearly half of the population belongs to household farms for which 30% of income is supported by livestock production.

Livestock production is largely concentrated in household farms (cattle 90%, horse 78%, camel 76%, and milk 80%). Peasant farms, registered farms with lands of long-term lease, have emerged after struggling for farm restructuring for a decade, but the revenue of average peasant farms seems to be narrowly viable. Some corporate farms reorganized voluntarily by farmers were revitalized by new financial support by the state and oil companies, but their future performance is unknown.

The study concluded that organizing livestock farms is important for the convenience of public support to enhance competitiveness in market economy and resource management, particularly at the periphery of settlements. Reorganizing household farms at the village level, formulation of service-oriented associations by peasant farms, development of social entrepreneurship corporation jointly with oil companies, and value addition of livestock products at the village level are suggested. Feasibility studies on such institutions, local-based wool and leather production, and new initiatives to graze rangelands are recommended.

Chapter 6 - The Sustainable Livelihoods (SL) framework is widely used for structuring qualitative studies of livelihood. This chapter tests the feasibility and value of doing a quantitative survey in the Taita Hills of Kenya using the SL framework. The framework was used as a structure both in the quantitative data collection and the statistical analysis of interrelationships of livelihood assets and outcomes. According to the survey, some of the biggest changes in the Taita Hills, as commonly perceived by farmers, are the declining soil fertility, decreasing number of livestock, new agricultural technologies, changes in the number of trees and species, and the increasing need of off-farm income. There is great variation in perceived changes, problems, adaptation strategies, coping strategies with unexpected expenditures, experimenting with new activities, and achieving desired livelihood outcomes. To study interrelationships of the different livelihood capital and livelihood outcomes several variations of multivariate regression analysis were used. The livelihood objectives that were considered by the farmers themselves as the most desirable in improving the standard of living were used to select the components of the livelihood outcome. Although the process is dynamic, with each household following a different time trajectory, it is expected that some links between the capital and the outcomes can be detected in cross-sectional data. The linear multivariate regression analysis showed that only financial capital could be directly related to the measured livelihood outcomes. But the variance accounted for is small, 26%. Analysing the components of outcome and capital separately reveal some patterns, and several of these challenge conventional expectations. However, the overall picture is one of much variation and individuality in livelihoods. The SL framework was found useful in the design of the survey. It focuses the survey instrument on the key elements, and it helps structure the presentation and analysis. However, it has some shortcomings when used with quantitative data. Much remains unexplained by a quantitative analysis using the sustainable livelihood SL framework. This may be due to the multiple livelihood strategies, selection of indicators, and/or the many intangible assets and social processes not described by the framework. However, despite the great variation, financial capital may be the most crucial asset in influencing the realization of livelihood outcomes. Increasing occupational multiplicity is a general trend in rural Africa. The increasing need of cash and the limited income from farm production is forcing farmers to look for off-farm jobs to supplement farm income.

Chapter 7 - The chapter explores the bio-physical and socio-economic determinants of adopting a diversified cropping system using a survey of 406 farmers located in 21 villages in three agro-ecological regions of Bangladesh. The computed value of the Herfindahl index of crop diversification confirms that farming system in Bangladesh is still relatively diverse despite four decades of thrust in the diffusion of "Green Revolution" technology package aimed at promoting modern rice monoculture. Although the gross value of output per hectare is significantly higher for specialized farms (i.e., farms concentrating on modern rice monoculture), the profits between specialized and diversified farms are similar because of significantly lower use rates of all variable inputs by the latter. Results from the Tobit model reveal that a host of factors, including input prices, significantly determines decision to adopt a diversified cropping system. Among the input prices, a rise in labour wage significantly promotes crop diversification, whereas a rise in the prices of fertilizers, pesticides and animal power services favour specialization towards modern rice monoculture. Crop diversification is positively associated with farm size. However, tenancy and availability of irrigation significantly favour specialization, implying that owner operators and farmers with no irrigation facilities are more likely to choose crop diversification. Both education and farming experience are significantly positively associated with crop diversification. Adoption of a diversified cropping system is significantly higher in regions endowed with developed infrastructure and relatively better soil fertility. Share of non-agricultural income also positively influence crop diversification. Therefore, adoption of a diversified cropping system could be promoted significantly by investing in rural infrastructure, farmer education, and soil conservation measures. Price policies aimed at improving labour wage and reducing fertilizers, pesticides, and animal power prices would also promote crop diversification. In addition, land reform policies focusing on delegating land ownership to landless and marginal farmers are noteworthy.

Chapter 8 - Cattle production in the humid lowland tropics has been criticised for its extensive management system, low productivity and high burden on the environment. Dual-purpose cattle production systems have been preferred by small and medium sized farms in such regions due to low capital and technical demands with low risk for farmers. Intensification of existing dual-purpose cattle farms has been recognized as a key target to reduce deforestation in Central America. However, the details of complex farming systems using seasonal cattle movement as well as diversified grazing lands with pasture and naturally regenerated trees are not well known. The objective of the chapter is to examine dual-purpose cattle production systems with regard to herd management, productivity and land use pattern. Seventy-four farms in the semi-humid region of central Nicaragua were seasonally monitored and relations with land use patterns were examined.

The study results showed that grazing lands were largely covered by tree cover (23% on average). Although cattle were frequently moved from and to the farms, stocking rates did not differ by season ($P<0.05$), suggesting that seasonal cattle movement towards more humid area is limited. The maintained stocking rates as well as significantly higher occurrences of calving ($P<0.05$) seemed to cause serious fodder shortage for cattle in the dry season, resulting in lower saleable milk production per cow and per hectare ($P<0.01$) and a tendency to have higher adult mortality rates. However, milk production per farm was not reduced in the dry season due to the larger number of calving. Fodder availability in the dry season significantly restricts potential of milk production. Smaller farms tend to have higher stocking rates and overgrazed, while some farms started introducing cut and carry system maintaining their pasture resources. Farms with higher financial resource improve their pasture, but shift their production types to steer rearing which requires less labour inputs.

The study concluded that shortage of dry season fodder, lack of calving control, and reluctance to increase payment for labour are the main obstacles for environmentally sound intensification of dual-purpose cattle farms. Further research is recommended on improvement of calving control, fodder availability in the dry season, the feasibility of pasture improvement, cattle movement and sharing system, and relations between farm manager and paid labour.

Chapter 9 - Due to ever-increasing population, water scarcity, and pressing issues of environmental sustainability, Asian rice farmers are under considerable pressure for sustainable increase in rice production by using less water. It is widely believed that an increase in the water use efficiency through integrated crop management holds promise of increased yields and water productivity. The so-called System of Rice Intensification (SRI) is attracting favorable attention of farmers and governments in Asia and elsewhere. It is assumed that a healthier and larger root system can be induced by using "SRI principles" in a water-limiting environment giving positive impacts on grain yield. The cultural practices that characterize SRI includes rapid and shallow transplanting of younger seedlings, at wider spacing and maintaining alternate wet and dry condition or preferably just moist conditions during the vegetative stage. This chapter reviews the biological mechanisms of water-saving agriculture and its relation to SRI cultural practices. It presents some research findings on the rice plant's adaptive trait which could be utilized to manage crops under limited water application. In addition, the details of on-farm studies carried out in some of the rice growing countries of Southeast Asia using a participatory action research approach are included in this chapter, which asserts the need for integrating science, people and policy makers for better and sustainable water management in Asia and elsewhere.

In: Agricultural Systems: Economics, Technology and Diversity ISBN 978-1-60692-025-1
Editor: Oliver W. Castalonge

Chapter 1

DRIVERS AND RESTRAINTS FOR ECONOMIC AND TECHNICAL EFFICIENCY IN DAIRY FARMS

Helena Hansson *
Department of Economics
Swedish University of Agricultural Sciences
Uppsala, Sweden

ABSTRACT

This chapter is a synthesis of five empirical studies (Hansson, 2007a; Hansson, 2007b; Hansson, 2007c; Hansson, 2008; Hansson and Öhlmér, 2008) which are all parts of a research project explaining reasons for inefficiency in dairy production. The analysis was conducted in light of four themes: the overall structure of inefficiency, the farm itself, differences in operational managerial practices and farmer managerial capacity.

The research project was based on a dataset of 507 dairy farms in Sweden. Farm efficiency scores, based on accounting data of each farm, were calculated with the data envelopment analysis. Factors hypothesized to explain differences between efficient and inefficient farms were collected with a questionnaire which was answered by about 320 farmers. Data were also obtained from a dairy cow recording scheme conducted by the Swedish Dairy Association. The factors hypothesized to explain reasons for inefficient production were analyzed with regression models.

The results showed that the links between efficiency and commonly used critical success factors are typically weak, implying that the critical success factors cannot predict efficient farming. The results also showed that farm size affect different efficiency measures in opposite ways, implying that the overall effect is unclear. Regarding strategic factors, considered in light of the third theme (the farm itself), the results showed that more diversified farms are more efficient, and also that larger fields and bunker or tower silos lead to more efficient farms. Farms situated in a social context, where the farmer can discuss milk production with someone on a regular basis, were more efficient. The results further showed that factors adjustable in the daily work typically do not affect farm efficiency to any larger extent. Finally, the results showed that farmers who are educated in agriculture and continually up-date their skills, e.g.,

* Helena.Hansson@ekon.slu.se

through study circles, generally run more efficient farms. The most fruitful insights into why some farms are inefficient were found in the themes the farm itself and farmer managerial capacity. The chapter ends with two suggestions for future research.

1. Introduction

Understanding why some dairy farms succeed in their farming, make profits and have flourishing businesses, whereas others do not is a pressing question. Knowledge of this is, of course, of great importance to the farmers themselves. If it is understood why some farmers succeed, less profitable farmers can learn from the successful ones and become more profitable. However, the question is pressing also for other agribusiness parties, such as the dairies which rely on the milk producing farms to get their raw material. Moreover, farms create opportunities for employment at the countryside and are therefore important to reach governmental goals such as successful rural development. Further, farms with grazing livestock are a prerequisite for biodiversity, and such farms are therefore necessary if we want to reach such societal goals. In this chapter I elaborate on the question of why some dairy farms in Sweden succeed in their business whereas others do not succeed. Indeed, this chapter is a synthesis of a four-year research project aiming at analyzing drivers and restraints for economic, technical and allocative efficiency at dairy farms in Sweden. The chapter focuses especially on how various aspects of technology, long-run strategic choices and farm management contribute to or prevent more efficient farms.

1.2. Background

The background for this research is firstly the substantial structural changes that are occurring at dairy farms in several countries at present. In Sweden, where the empirical work in this research was based, the number of dairy farms has decreased by 67%, from 25,921 farms in 1990 to 8,548 farms in 2005 (Statistics Sweden 2006). During the same time the average herd size more than doubled, from 22 to 46 cows. Meanwhile, the amount of milk produced decreased by only 8% (Statistics Sweden 2006). These figures suggest that Swedish dairy farms are becoming fewer and larger. Owners of small farms quit farming, moved into other production lines or merged with other farms. Similar structural changes can be traced also in other north European countries. For instance, in Denmark the number of dairy farms decreased by 39% between 1997 and 2003 (Statistics Sweden, 2002; 2006) whereas their average herd size grew by 44% to 75 cows (Statistics Sweden, 2002; 2006). This trend, with fewer and larger dairy farms, is also reported for the U.S. (Tauer and Mishra, 2006; MacDonald et al. 2007).

Secondly, while the farms get fewer and larger, the Mid Term Review, which is part of the Common Agricultural Policy (CAP) reform in Europe, changes the support levels so that they are no longer proportional to the production level. This implies that the market situation of the farms change, which is likely to cause new managerial challenges at the farms. Indeed, nowadays the world market prices determine the output prices of milk, a situation that implies that, e.g., the Chinese demand for cheese and milk powder determine the output prices of milk in countries such as Sweden.

Thirdly, a key background factor for the research presented in this chapter is that several international efficiency studies on dairy and related livestock farms show some important and unambiguous results: the profits at these farms would increase if their inputs or outputs were allocated more optimally. In efficiency studies, each individual firm is compared to the best practice in the sample at hand. Thus, the results show that if all farms were as efficient as the best ones, efficiency could increase. In Sweden, Heshmati and Kumbhakar (1994) found that outputs could increase on average nearly 20% if the dairy farmers produced as much as they could, given their inputs. They studied four panels of farms during 1976 - 1988, excluding 1985, and found levels of technical output efficiency of on average between 81% and 83%. Lawson et al. (2004) studied technical output efficiency in Danish dairy farms, and found that they could increase their outputs by about 5%. Bravo-Ureta and Rieger (1991) found similar levels of technical input efficiency as Heshmati and Kumbhakar (1994) for dairy farms in New England. In that study the authors also estimated farm level economic and allocative input efficiency, finding average levels of 70.2% and 84.6% respectively. Their results imply that the New England farms could reduce their costs by almost 30% if they allocated their inputs in a cost minimizing way and did not over use any of their inputs.

A common feature of all these referred studies is that they base their conclusions on stochastic frontier analyses (SFA). An alternative method is the data envelopment analysis (DEA). This method was used in a study by Tauer (1993) who reported average technical input efficiency of 85% and allocative input efficiency of 70% for New York dairy farms. DEA was also used in a study by Oude Lansink et al. (2002) to assess the technical input efficiency of Finnish livestock farms. They found an average efficiency level of 69% for conventional farms and 93% for organic farms. Although the literature shows that dairy farms, on average, can increase their profits, it is not possible to argue in what studies, or in what countries, the more efficient farms are found. Differences in research method (Coelli et al. 2005) cause at least some of the differences. Furthermore, differences in data and variable specification may cause differences in average efficiency levels. However, because both SFA and DEA estimate an efficient frontier based on the results of the most successful farms in the sample, low average levels of farm level efficiency can be argued to show that there is high variation in the efficiency levels. Consequently, some farms are much more efficient than other farms, and therefore there are large possibilities for the less efficient ones to improve their efficiency, e.g., by adapting to the practices used by the most efficient farms.

1.3. Problem and Aim

The first two developments outlined above change the environment in which the Swedish dairy farmers are working. This is likely to give rise to new challenges facing the dairy farmer as business owners, problems which they had not previously been particularly familiar with. These farms have traditionally been managed by only the farmer family and they have also relied mostly on equity. However, now when the farms become increasingly larger this situation is unlikely to persist. Instead, the farms are likely to start having employees and a higher debt-equity ratio. Sensitivity to slacks due to inefficient production is therefore likely to become more severe. In this setting, the management aspect of dairy farms will become more important. These developments are important not only because the business style of the dairy farmers needs to change, but also the bulk of knowledge of their advisors and business partners will have to develop so that they can meet the requirements of the new dairy farm environment. The third development outlined above shows evidence that dairy farms can improve their profits by becoming more efficient. This naturally raises questions such as: What are the characteristics of successful dairy farms? What drives successful dairy farming? What restrains it? Knowledge of what constitutes driving and restraining forces for successful dairy farming is essential both to assist farms that are not as efficient as they could be and to assist even the efficient farms to become more sustainable in a changing surrounding world.

As stressed earlier, a four year empirical research project has aimed at analyzing the problems sketched above. Based on the five separate empirical studies conducted within the research project, all with the common over-all aim of analyzing what drives and restrains efficient production in dairy farms, the aim of this chapter is to synthesize the main findings of the project in light of previous knowledge, to conclude on concrete actions to strengthen the farms and help them become more efficient. Furthermore, this chapter also aims at outlining ideas for future research within this area. A synthesis of the main findings contributes by giving a deeper and more comprehensive picture of what drives and restrains efficiency in dairy farming, than what is obtained from the individual studies that have emerged from the research project.

This chapter is organized as follows: Section 2 reviews previous literature and forms a basis of four main themes on which the research project relied. Section 3 discusses efficiency as a way of analyzing farm performance. Section 4 discusses methodological issues dealt within the research project. Section 5 presents and discusses the data used in the research project. Section 6 summarizes and discusses the main findings of the research project. Section 7 finally ends the chapter with conclusion and outline ideas for future research.

2. Previous Literature

Classical literature on firm behaviour offers several dimensions to explain how firms work both internally as well as in their environments and how they act to be sustainable. Although not necessarily written to explain why production is inefficient, the classical literature offers a background to understand why firm behaviour deviates from what is rational and efficient. Simon[1] (1997) described behaviour in firms principally by discussing how decisions are made in organizations. He pointed out that time, abilities, values and knowledge of individuals constrain decision making such that rationality is constrained. Furthermore, Simon (1997) maintained that when individuals have their attention set in some direction, this tends to be relatively persistent over time. Arguably, these characteristics should influence firm behaviour. Cyert and March (1963) expressed four concepts that they argued to be fundamental to understand firms: Quasi resolution to conflict, avoidance of uncertainty, problemistic search and organizational learning, which all add to understand firm behaviour. By quasi resolution to conflicts Cyert and March (1963) meant that because there are always different goals in firms that are in conflict with each other, e.g., large market shares and profits, conflicts arise which are attended to sequentially and by certain decision rules. Avoidance of uncertainty is the behaviour that arises because firms have a short-sighted behaviour, in that they act by responding to short-run problems rather than through long-run

[1] Reprinted version of Simon's classical book, first printed in 1945.

strategies. Short-run problems are handled by e.g., setting standards with the environment to reduce or remove uncertainty. By problemistic search, Cyert and March (1963) meant that search is stimulated by an awareness of a problem and that search aims to find a solution to the problem. Finally, the organization learns when the individuals within it adjust to new situations. Pfeffer and Salancik (1978) offered a resource dependence perspective to how firms act and develop. In this perspective, firms are understood in the context in which they exist. Other firms or individuals drive or restrain the behaviour of a particular firm, when the firm responds to the environment. Firms thus exist in interdependence with other firms and individuals, which determine their development.

The resource based view (RBV) is another stream of literature which generates fruitful insights into why some firms are efficient and successful. Barney (1991) and Barney et al. (2001) presented RBV as a theory that describes the sustained competitive advantage, i.e., success, of firms from the resources and capabilities the firm controls. Resources in this sense are a wide concept, covering both tangible and intangible assets, including the management skills of the people in the firm, its organization and information. The RBV argues that possession of unique resources that are difficult to imitate by other firms leads to sustained competitive advantage (Barney, 1991). In the context of this chapter, the RBV offers a fruitful framework in that it suggests that reasons for successful farming lie in the resources controlled by the farm. This can be compared to e.g., Porter (1980) who described strategic management - and therefore in the continuation keys for success - as adjusting properly to the surrounding environment, such as the behaviour of competitors. If the studied resources are controlled by the firm, they should at least to some extent be adjustable by the firm. Thus, less efficient farms can then adjust their resources towards the better.

2.1. Suggestions to Why Farms Are Not Efficient

Literature on farm management emphasize the importance of goals and values of farmers as motivations for being a farmer and as explanations for why farmers do not always maximize profits. For instance, as argued in Boehlje and Eidman (1984), students of agriculture need to recognize the specific characteristics of farms such as the closeness between the family and the farm. Normally it is not possible to separate the business from the family and the leisure time. In this setting, Boehlje and Eidman (1984) argued that other goals than profit maximization may more easily come to the farmer's mind, for instance controlling a larger farm or having a low debt-equity ratio.

Differences in managerial capacity are stressed in the literature as conceivable reasons for why farms differ in level of efficiency (Rougoor et al. 1998; Wilson et al. 1998; Nuthal 2001). Rougoor et al. (1998) developed a framework which showed the link between managerial capacity, in terms of personal aspects of the farmer and the decision making process, and farm level efficiency. Beyond personal characteristics such as the age, education and experience of the farmer, economic psychological aspects are also interesting in explaining farmer behaviour. For instance, the goal of the farmer may influence farm efficiency. Gasson (1973) identified four goals of farmers: instrumental, expressive, social and intrinsic. Of these, social and expressive goals are likely to affect farm performance negatively, because focus is on gaining prestige as well as being creative and original. Further, the perception of the farmer, i.e., how he or she sees the world (e.g. Hogart, 2001) is likely to influence farm efficiency. Locus of control is another interesting aspect, which indicates how the individual perceives his or her ability to influence his or her decisions. Daft (2003) stressed internal locus of control, i.e., a situation where the individual perceive his or her own decisions to affect the situation, as an important personality trait of entrepreneurs. The decision making process of farmers was described by Öhlmér et al. (1998). They described how farmers make decisions, rather than how they should make decisions. Based on 18 case studies, the decision making process was suggested by Öhlmér et al. (1998) to consist of four phases: problem detection, problem definition, analysis and choice, and implementation. Moreover, Öhlmér et al. (1998) suggested that each phase consists of four sub processes: searching for information and paying attention, planning, evaluating and choosing, and bearing responsibility. Decision-making should thus be viewed as a matrix rather than linear steps following each other.

In the managerial situation, information handling is recognized as an essential part to detect and solve problems: Klein et al. (2005) meant that interpretation of information from the environment is a prerequisite to influence one's situation. Two mental systems can be at work when individuals interpret and act on information: the tacit and the deliberate systems (Hogart, 2001). When the tacit system is at work, information is interpreted with the aid of previous experiences and values. In the deliberate system, careful calculations play a prominent role. Because the deliberate system is costly, Hogart (2001) argued that it is plausible to assume that the experience-based, tacit system is the default system and that the deliberate system is used more scarcely. The tacit system is developed with experience, which works to strengthen the individual's mental models (Klein et al. 2005). However, if experience is acquired in a wicked environment, it will cause the tacit system to produce dysfunctional behaviour (Hogarth, 2001).

Strategic management literature is also interesting when trying to understand why some farms are efficient whereas others are not. Strategic management is the long-term process where aspects that are important for the performance of the farm are taken into consideration. Several factors that are decided on in the strategic management are difficult to change in the short run, and therefore these factors lay a basis for the farm and its performance. Strategic questions were emphasized by Harling (1992) who found that successful farmers tend to think in terms of strategic management to a greater extent compared to their less successful counterparts. Lee et al. (1999) described three environments that influence firm strategy: external, operational and internal. The firm has no control over the external environment, which corresponds to macro-economic conditions. The operational environment describes the market situation, and Lee et al. (1999) maintained that the firm may have some control over this environment. The internal environment contains resources under direct control of the firm. I argue that yet another dimension should be added: the micro-social environment which consists of e.g., the farm family. The farm family is often stated as being of major importance for the farm business (Gasson et al. 1988; Harling and Quail 1990). The aim of adding the micro-social environment is consequently to describe potential driving and restraining forces within the nearest social environment in which the farm exists.

Differences in routines adjustable in the daily farm work are also interesting when trying to understand differences in farm efficiency. For instance, routines associated with animal health, breeding and feeding practices can be, if not totally changed, at least improved in the daily work towards a situation that is better from an efficiency point of view. Previous literature has found that such practices can affect the production at the farm. For example,

Steinbock et al. (2003) found that a too low age of the heifer at her first calving can cause problems such as uneasy calving and still births. Longer lactations, an effect of longer calving intervals, was found by Bertilsson et al. (1997) to increase milk yield. Further, Bertilsson et al. (1997) also stressed that longer calving intervals may reduce udder problems arising from drying off cows that are still high yielding. Concerning feedstuff, it is interesting to note that hay is generally considered to have a higher value protein content compared with silage (Shingfield et al., 2002). Therefore, feeding hay should increase milk quality.

Apart from the descriptive understanding of the management at the farm, analyses of managerial devices used by farmers to manage their farms are an essential part to understand why farms are not as efficient as they can be. The managerial device critical success factors were found by Huirne et al. (1997) to be used and considered as important by dairy farmers. These factors are defined as indicators aimed at pointing out successful business management (Rockart, 1979). Management's critical success factors consist of a broad range of factors, both financial and non-financial and can be used both for comparisons over time as well as between farms.

2.2. Efficiency Studies and Analyses of Causes of Inefficiency

Efficiency studies as they are known today began with the seminal paper by Farrell (1957) who gave precise definitions of efficiency in firms. Efficiency is defined from two perspectives: the input-oriented (cost) perspective and the output-oriented (revenue) perspective. In both perspectives, three main efficiency scores are defined: economic, technical and allocative efficiencies. Methodology to estimate efficiency was established by Aigner et al. (1977) and Meeusen and van den Broeck (1977) who introduced an econometric way of estimating efficiency, as well as by Charnes et al. (1978) who introduced mathematical programming tool to calculate the efficiency scores. Since then, numerous applications on firm level efficiency have been made in the empirical literature, with applications to a broad array of firms. The most commonly considered efficiency score is the technical efficiency score. Possibly because it can be difficult to get information about prices, usually only this type of efficiency score is estimated, or from only one perspective.

In the efficiency literature, it is common to evaluate how some explanatory variables affect efficiency. Examples in the agricultural literature include Bravo-Ureta and Rieger (1991), Sharma et al. (1999), Coelli et al. (2002), Iraizoz et al. (2003), Helfand and Levine (2004), Lawson et al. (2004), Latruffe et al. (2005). Farm size is a recursive determinant of efficiency. In light of the development in Sweden where farms grow fast, it is urgent to understand how farm size affects efficiency. Some authors have found a relationship between the physical size of the farm and efficiency. Most of them have made their applications to other farms than dairy (for example Helfand and Levine 2004; Iráizoz et al. 2003 and Sharma et al. 1999), but Bravo-Ureta and Rieger (1991) and Alvarez and Arias (2004) studied dairy farms. In most of these studies the relationship is positive; however the results are not consistent clear across studies.

A common characteristic of the literature is that the analyses of the determinants of efficiency are conducted only in a broad outline. Even though differences in management are widely recognized as essential when trying to explain differences in farm efficiency, previous literature that has statistically tested what explains efficiency has generally only broadly covered these factors. Normally factors such as the age or the education of the farmer are included as an approximation of the management factor. Only a few studies exist that have included aspects of strategic management and managerial systems in light of farm level efficiency. For example, Wilson et al. (2001) included business goal and the number of information sources used in the decision making process as explanatory variables for technical inefficiencies at wheat farms in England. Trip et al. (2002) modelled technical inefficiencies as dependent on the goals of the manager and the quality of the decision making process, in terms of planning, data recording and evaluation. Aspects of strategic management were studied in light of farm level efficiency in de Koeijer et al. (2003) and Ondersteijen et al. (2003).

2.3. Starting Points to Study Drivers and Restraints for Efficient Production

Based on the literature review, four broad themes emerged as fruitful starting points to understand what drives and what restrains efficiency in dairy farms and consequently to understand how the farms can develop. First, we need to better understand the broad structure of inefficiency in the dairy production. This triggers research questions such as: What efficiency perspective and score indicate the more problematic part and consequently the more important part for the farmers to focus on if they are to become more efficient? How good are the dairy farm management's critical success factors at indicating farm efficiency? How does farm size affect efficiency? The experienced need of further studies of the relationship between farm efficiency and farm size needs a comment. Previous literature (e.g. Bravo-Ureta and Rieger, 1991; Alvarez and Arias, 2004) has indeed studied this relationship at dairy farms before, but these results are not easily translated to Swedish dairy farms, because of differences in technology, climate and business culture. Further, previous literature does not show an unambiguous relation between farm efficiency and size.

Second, the effect of characteristics describing the farm itself needs to be understood. This involves understanding how aspects decided in the long-run strategic decision making affect efficiency. Strategy factors are factors that should be under consideration especially when a farmer decides to stay in business and expand the farm or when new farms are considered. These factors are difficult to change in the short-run, and therefore lay the long-run basis of the farm. In particular, how determinants of the external-operational and internal environments (Lee et al. 1999) as well as the micro-social environment affect farm performance is essential to analyze.

Third, the effect of factors adjustable in the short-run managerial practices are interesting to study because they can, if not be completely changed, at least adjusted in the short run and therefore facilitate more efficient production. These include factors describing animal health, breeding and feeding practices.

Fourth, finally, further and deeper analyses of the farmer managerial capacity, in e.g., terms of farmer characteristics and decision making as suggested by Rougoor et al. (1998) are vital to understand how farmers can be supported to become better at managing their farms.

Table 1 shows a list of refined research questions and their study object(s), based on the literature review.

Table 1. The research questions and their study object(s) covered in the research project

Themes and research questions	Study object(s)
The structure of inefficiency	
What efficiency score is the more problematic? Are there any links between management's critical success factors and efficiency ? Are larger farms more efficient?	The overall structure of efficiency
The farm itself	
What are the effects of factors decided in the long-run strategic decision making?	The external-operational environment of the farm, the farm as such (its internal environment) and the micro-social environment.
Operational managerial practices	
What are the efficiency effects of factors that are adjustable in the short-run operational work?	Animal health, breeding and feeding practices.
The farmer managerial capacity	
What are the effects of the managerial capacity of the farmer?	The personal characteristics the farmer and management systems that facilitate decision making.

3. Efficiency as a Way of Analyzing Farm Performance

As already mentioned in the literature review, efficiency studies normally builds on the framework developed by Farrell (1957), which defines efficiency from an input and output-oriented perspective. The input oriented perspective focuses on the cost side, i.e., how much inputs or costs can be reduced, while the given amounts of outputs are still produced. The output-oriented perspective, on the other hand, focuses on the revenue side of the firm and measures how much output or revenue can be increased while the given amounts of inputs are used. Three main types of efficiency scores are defined in each perspective: economic, technical and allocative. Efficiency is defined in the 0 - 1 region, where a score of 1 indicates full efficiency. The efficiency score subtracted from 1 defines the farm's corresponding inefficiency. Technical inefficiency measures how much inputs can be reduced given the level of outputs (input-oriented efficiency) or how much outputs can be increased given the level of inputs (output-oriented efficiency). Allocative inefficiency measures how much costs can decrease if the combination of inputs was optimal according to prices (input-oriented efficiency) or how much revenues can increase if the combination of outputs was optimal according to prices (output-oriented efficiency). Economic inefficiency can be said to measure overall efficiency, in that it is a combined measure of both technical and allocative efficiencies. If constant return to scale is assumed or the actual case, the technical efficiency score is the same regardless of perspective: however this is not the case for economic and allocative efficiencies.

It is not self evident that successful farms should be defined according to efficiency scores. Other ways of defining successful farms could be traditional financial ratio analyses and cost-revenue analyses. However, there is a major advantage of the efficiency approach: it evaluates the farms in a comprehensive way, in that all inputs and all outputs are considered at the same time (Coelli 1995). The financial ratio analyses and the cost-revenue analyses on the other hand, compare only two aspects at a time. A critique that can be addressed against defining success according to an efficiency score is that the efficiency analysis assumes that it is desirable to improve the cost and revenue situation at the farm. If the farmer has other goals than the instrumental ones, he or she may succeed in fulfilling these goals even though instrumental goals are not fulfilled. Nevertheless, I argue that to be sustainable in the long run, farmers need to at least consider their costs and revenues, regardless of what their primary reason for being a farmer is. As a consequence, the efficiency analysis is a convenient way of assessing farm level success.

3.1. What Is Efficiency and Inefficiency?

To describe efficiency and inefficiency, we need to start by defining the technology of the firm. For both the input and the output perspective the firm technology T can be represented by distance functions, which can be described as collections of sets. The inputs used by firm *i* are defined as the vector $x_i = (x^1,.....,x^h) \in \Re_+^h$, where *x* defines each input and h is the number of inputs. The outputs of firm *i* which are produced by the inputs, are defined by the vectors $y_i = (y^1,.....,y^m) \in \Re_+^m$, where *y* defines each output and m is the number of outputs. Production occurs according to the technology $T = \{(x_i, y_i) \in \Re_+^{h+m} \mid x_i \ \ can \ \ produce \ \ y_i)\}$. The properties imposed on the technology are nonemptiness, closedness, convexity and free disposability of inputs and outputs.

3.1.1. Input-Oriented Perspective

Under the input-oriented perspective, knowledge of the fully efficient isoquant makes it possible to describe the technical efficiency of the firms in the sample. If the isocost line is also known, allocative and economic efficiencies can be estimated. An economically input-efficient firm uses the smallest and cheapest combination of inputs to produce a given output. If single output and two inputs are assumed, the efficiency indices can be illustrated graphically as in Figure 1.

The isoquant YY' represents the technically efficient way to produce the given input Y. The economically optimal point is at Q', the tangency point between the isoquant YY' and the isocost line, PP', where the technical rate of substitution between the two inputs equals the economic rate of substitution. If an inefficient firm is represented by the point R and produces the amount Y, its economic efficiency is measured as 0R'/0R, and the corresponding inefficiency is 1 – 0R'/0R which is interpreted as a potential cost reduction. Economic inefficiency can consist of two parts: technical inefficiency and allocative inefficiency.

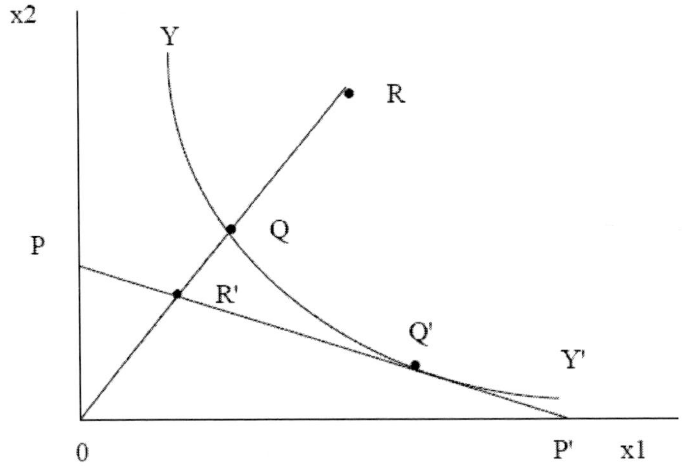

Figure 1. Technical, allocative and economic input efficiency.

The technical efficiency is measured as 0Q/0R and the allocative efficiency is measured as 0R'/0Q. The economic efficiency is then recognized as the product of technical and allocative efficiency.

3.1.2. Output-Oriented Perspective

In the output-oriented perspective, efficiency is evaluated keeping inputs constant. Knowledge of the fully efficient production possibility curve as well as the isorevenue line makes it possible to measure and interpret the economic output efficiency. The output oriented perspective is shown in Figure 2.

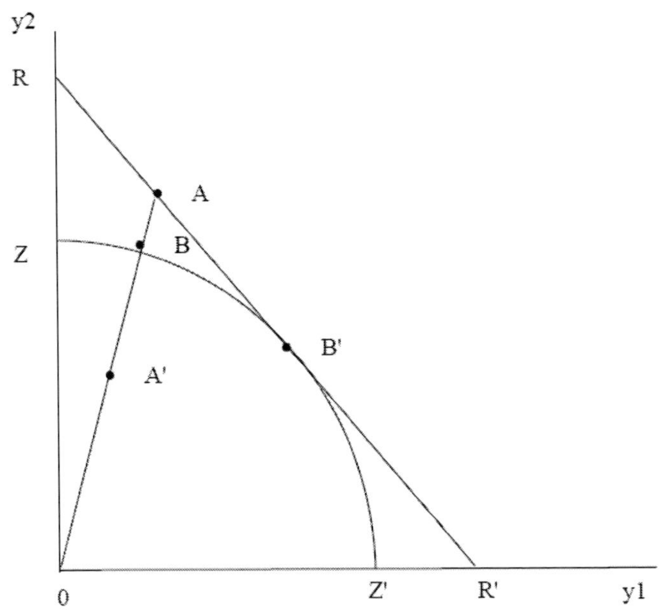

Figure 2. Technical, allocative and economic output efficiency.

The production possibility curve is represented by the curve ZZ' in Figure 2, which represents technically efficient combinations of production of outputs y1 and y2. The economically efficient point is B' where the marginal rate of product transformation equals the slope of the isorevenue line RR'. Consider a firm situated at point A'. Its economic output efficiency is 0A'/0A. The corresponding inefficiency is 1 - 0A'/0A, which is interpreted as a potential increase in profits. Technical efficiency is represented by 0A'/0B and the allocative efficiency is 0B/0A.

3.1.3. Summary of the Efficiency Scores

According to the framework outlined above, there are six major efficiency scores. In the short run, one may argue that some inputs are fixed. In that case, the farmer can reduce only the variable inputs to their optimal levels given the level of fixed inputs (Coelli et al. 2005). Consequently, nine major efficiency scores exist. These are, together with their economic interpretations, summarized in Table 2.

Table 2. The efficiency scores and their economic interpretation

Measure of efficiency	Economic interpretation
Long-run economic input efficiency	Produce a given set of outputs using the smallest and cheapest set of inputs, in the long run.
Long-run technical input efficiency	Produce a given set of outputs using the smallest set of inputs, in the long run.
Long-run allocative input efficiency	Combine inputs in cost-minimizing proportions, in the long run.
Short-run economic input efficiency	Produce a given set of outputs using the smallest and cheapest set of inputs, in the short run.
Short-run technical input efficiency	Produce a given set of outputs using the smallest set of inputs, in the short run.
Short-run allocative input efficiency	Combine inputs in cost-minimizing proportions, in the short run.
Economic output efficiency	Produce the maximal set of outputs, given the set of inputs, while maximizing revenue.
Technical output efficiency	Produce the maximal set of outputs, given the set of inputs.
Allocative output efficiency	Combine outputs in revenue-maximizing proportions.

4. METHODOLOGICAL CONSIDERATIONS IN THE RESEARCH PROJECT

The research project synthesized in this chapter has used a quantitative approach throughout the whole project. Typically, efficiency scores have been assessed in a first step, and in a second step regression analyses have been used to determine the effect of possible drivers or restraints on farm efficiency.

4.1. Method to Estimate Farm Efficiency

Data envelopment analysis (DEA) (Charnes et al. 1978) was used to estimate farm efficiency. DEA is one of the two most popular techniques to measure farm level efficiency. The other one is the stochastic frontier approach, SFA (Aigner et al. 1977; Meeusen and van den Broeck 1977). The DEA uses mathematical linear programming methods, whereas the SFA uses econometric methods. Both methods are empirical approaches: both DEA and SFA base their efficiency assessments on the best practice in the sample at hand, so that the best farms define the efficient frontier. The remaining farms get efficiency scores according to their relative position to the efficient frontier. This has implications for research based on either one of these two methods, such as the research synthesized in this chapter, because it implies that what is studied is how the least efficient farms can become as efficient as the best practice farms. However, it is of course possible that even the best practice farms could be more efficient if their production processes were theoretically optimized.

The quality of the data, the appropriateness of various functional forms, and the possibility of making behavioural assumptions influence the relative appropriateness of DEA and SFA. For example, the DEA approach, compared with the SFA requires no specific functional form to be selected, neither are any behavioural assumptions needed. However, DEA is a deterministic approach, meaning that it does not account for noise in the data. All deviations from the frontier are therefore accounted for as inefficiencies. The DEA efficiency scores are, therefore, likely to be sensitive to random errors. The SFA, on the other hand, accounts for random errors and has the advantage of making inference possible (see for example Coelli et al. 2005). However, SFA is likely to be sensitive to the choice of functional form.

Empirical comparisons between SFA and DEA have been conducted during the last years to evaluate if the efficiency results differ depending on the method chosen. Examples from both inside and outside the agricultural sector include Resti, 1997, Sharma et al. 1999, Coelli and Perleman, 1999, Reinhard et al. 2000, Iraizoz et al. 2003 and Cullinane et al. 2006. The general results obtained in the comparisons are that DEA and SFA produce similar average efficiency scores or that the DEA scores are slightly lower than the SFA scores. Further, there are generally no significant differences in the rankings of efficient farms, which means that relative efficiency is given to the same farms in both methods.

As stressed above, DEA was used in the research project synthesized in this chapter. The reason was that some advantages such as its ease of decomposing economic efficiency into its allocative and technical parts, as well as no requirements of specification of functional form were experienced with DEA.

4.1.1. DEA Equations

To estimate DEA efficiency scores, assume a sample of N individual farms. Together the farms use the input matrix X to produce the output matrix Y. Let x_i and y_i represent the input and output matrices of each individual farm i. Each farm faces a cost-minimizing input combination represented by x_i^*, a revenue-maximizing output combination y_i^* and vectors of input- and output-prices represented by w_i and p_i respectively. Further, in the short run the input matrix X_v corresponds to the variable inputs of all farms and the input matrix X_f corresponds to the fixed inputs of all farms. Each individual farm faces both variable and fixed input matrices, denoted x_{vi} and x_{fi}. The cost-minimizing input combination of variable inputs for the ith farm is x_{vi}^* and the corresponding price vector is w_{vi}.

For the ith farmer, the long-run minimum cost, which is the cost that would occur if the farm was operating at its cost-minimizing level, is obtained by solving the following linear programme:

$$\begin{aligned} & \min_{\lambda, x_i^*} w_i' x_i^* \\ \textit{subject to} \quad & -y_i + Y\lambda \geq 0, \\ & x_i^* - X\lambda \geq 0, \\ & N1'\lambda = 1, \\ & \lambda \geq 0 \end{aligned} \tag{1}$$

$Y\lambda$ and $X\lambda$ are the efficient projections on the frontier. $N1'\lambda = 1$ is a constraint ensuring variable returns to scale. Long-run economic input-oriented efficiency is obtained in a second step by comparing the minimized cost calculated by equation 3 to the actual cost:

$$EI_i = \frac{w_i' x_i^*}{w_i' x_i} \tag{2}$$

The short-run minimum cost is obtained by minimizing the cost of the variable inputs, which is done by solving the following program:

$$\begin{aligned} & \min_{\lambda, x_i^*} w_{vi}' x_{vi}^* \\ \textit{subject to} \quad & -y_i + Y\lambda \geq 0, \\ & x_{vi}^* - X_v\lambda \geq 0, \\ & x_{fi} - X_f\lambda \geq 0, \\ & N1'\lambda = 1, \\ & \lambda \geq 0 \end{aligned} \tag{3}$$

Short-run economic efficiency is then found by comparing the short-run minimum cost to the actual short-run cost:

$$EI_{si} = \frac{w_{vi}'x_{vi}^*}{w_{vi}'x_{vi}} \tag{4}$$

The long-run technical input efficiency is calculated by solving the following linear program:

$$\begin{aligned} & \min_{\theta_i,\lambda} \theta_i \\ \textit{subject to} \quad & -y_i + Y\lambda \geq 0, \\ & \theta_i x_i - X\lambda \geq 0, \\ & N1'\lambda = 1, \\ & \lambda \geq 0, \\ & \theta_i \in (0,1] \end{aligned} \tag{5}$$

where θ_i is the *i*th farmer's level of long-run technical efficiency.

The short-run technical efficiency is solved by the following program:

$$\begin{aligned} & \min_{\theta_{si},\lambda} \theta_{si} \\ \textit{subject to} \quad & -y_i + Y\lambda \geq 0, \\ & \theta_{si}\, x_{vi} - X_v\lambda \geq 0, \\ & x_{fi} - X_f\lambda \geq 0, \\ & N1'\lambda = 1, \\ & \lambda \geq 0, \\ & \theta_{si} \in (0,1] \end{aligned} \tag{6}$$

where θ_{si} is the short run technical input efficiency of the *i*th farm.

Both long- and short-run allocative input efficiencies are calculated residually:

$$AI_i = \frac{EI_i}{\theta_i} \tag{7}$$

where AI_i is the long-run allocative input efficiency for farm *i*, and

$$AI_{si} = \frac{EI_{si}}{\theta_{si}} \tag{8}$$

where AI_{si} is the short-run allocative input efficiency of farm *i*.

Economic output efficiency is calculated by first solving the following linear program:

$$\begin{aligned} & \max_{\lambda, y_i^*} p_i' y_i^* \\ \textit{subject to} \quad & -y_i^* + Y\lambda \geq 0 \\ & x_i - X\lambda \geq 0 \\ & N1'\lambda = 1 \\ & \lambda \geq 0 \end{aligned} \tag{9}$$

Equation 9 calculates the maximal revenue that the farm can receive if outputs are combined in their optimal way. Economic output efficiency, EO_i for the *i*th farm, is then solved by the following equation:

$$EO_i = \frac{p_i' y_i}{p_i' y_i^*} \tag{10}$$

where the actual revenue is compared to the maximal revenue that would be obtained if inpts were combined in the optimal way.

Technical output efficiency is solved by the following program:

$$\begin{aligned} & \max_{\phi_i, \lambda} \phi_i \\ \textit{subject to} \quad & -\phi_i y_i + Y\lambda \geq 0, \\ & x_i - X\lambda \geq 0, \\ & N1'\lambda = 1 \\ & \lambda \geq 0 \\ & 1 \leq \phi_i < \infty \end{aligned} \tag{11}$$

where the inverse of ϕ_i, $\frac{1}{\phi_i}$ is the technical output efficiency of farm *i*. If the farm operates under constant returns to scale, this is the same as long run technical input efficiency.

Allocative output efficiency, AO_i, is calculated residually as in the input-oriented cases:

$$AO_i = \frac{EO_i}{\frac{1}{\phi_i}} \tag{12}$$

If input and output prices are assumed to be given, the calculations of economic efficiency can be facilitated by reducing equations 1 and 2, 3 and 4, and 9 and 10 to the same principal form as the equations for the technical efficiency scores.

4.1.2. Method to Assess the Effect of Drivers and Restraints on Farm Efficiency

The literature describes several approaches to assess the effect of factors hypothesized to explain efficiency (see e.g., Coelli et al. 2005 for a review). The method chosen in this step depends to a large extent on the method chosen to assess the farm level efficiency scores. In the case of the research project synthesized in this chapter, DEA has been used, as motivated above. To determine the effect of factors explaining efficiency, multiple regressions were undertaken in a second step. DEA assesses full efficiency to at least some farms in the sample, which are considered the best practice in the sample. The remaining observations get efficiency scores in comparison with this best practice. Consequently, the efficiency scores cannot have a higher value than one; and an interesting question is whether the data are censored or truncated. Data are censored if, for some reason a maximal or minimal value is put on a variable (see e.g., Kennedy 1998).

This means that the dependent variable does not necessarily have the right value for those observations which have the censored value. This is what happens if, for instance, ability of students is measured by their test scores. All students getting the maximal score are not necessarily equally competent. Data are truncated if, for instance, all observations above a certain value are excluded (see e.g., Kennedy, 1998) and there exist neither dependent nor independent variables for these observations.

An example is when data about wages above a certain limit are missing. In the DEA case not all farms scoring one are exactly equally efficient, because the reductions in the number of observations as well as increases in the number of variables causes the efficiency scores to be biased against one. Consequently, if the method allowed for the efficiency scores to be greater than one, some of the farms would get higher efficiency scores than one. Based on this, DEA efficiency scores can be considered as censored. The censored nature of the efficiency scores has to be taken into consideration when assessing the impact of the factors explaining efficiency. In the literature (e.g. Sharma et al., 1999; Coelli et al. 2002), this is normally done by using the tobit regression, which allows censored dependent variables.

Tobit regression was used also in the research project synthesized in this chapter. Simpler alternatives, like the ordinary least square regression, may cause predicted values of efficiency to exceed one, which is not desirable. On a few occasions in this research project, however, logistic regressions were used instead, because that modelled better what was demanded.

The tobit model can be described as follows:

$$y_i^* = \sum_i \beta_i x_i + \varepsilon_i, \quad i = 1,2,....N,$$

$$y_i = 1 \quad if \quad y_i^* \geq 1 \tag{13}$$

$$y_i = y_i^* \quad if \quad y_i^* < 1$$

where $\varepsilon \sim N(0, \sigma^2)$ and the β_i are the parameters for the explanatory variables. (For a review of tobit regression, see for example Hayashi, 2000).

The method used in this research project can thus be described as a two-stage process, commonly used in the literature, where the efficiency scores are determined in the first stage and the effects of factors hypothesized to determine efficiency are determined in a second stage regression. In a recent paper by Simar and Wilson (2007) this approach was criticized. Their criticism is based on the fact that DEA produces biased efficiency scores in small samples and because the explanatory variables in the second stage regression are correlated to the inputs and outputs used to construct the DEA efficiency scores. To overcome these problems, Simar and Wilson (2007) suggest two bootstrap algorithms, of which the second algorithm (the double loop) corrects for both problems. Afonso and Aubyn (2006) compared these bootstrap algorithms with the traditional two-step approach in an empirical setting. They found that the estimated coefficients and significance levels were very similar in all three cases. This questions the value of extra computational burden cased by the bootstrap algorithms. Furthermore, and in the case of the research project described in this chapter more important, following Simar and Wilson (2007) would mean that it would not have been possible to use large parts of the data. Throughout the research project presented in this chapter, data have been pooled from different sources (see Section 5). This gave missing values for some explanatory variables of several observations. The data on inputs and outputs used to construct the efficiency scores are complete, but the data on factors hypothesized to affect efficiency have missing variables in several cases. If the two stage approach was used, all data on inputs and outputs could be used to calculate the first stage efficiency scores. In the second stage the observations where there were no missing values on the explanatory variables were used. Following the suggestion by Simar and Wilson (2007) would have meant that large parts of the data on inputs and outputs could not be used, because all observations needed to be involved in the entire bootstrapping process.

5. Data

Dairy farms in Sweden are generally operated at rather small farms, which are often owned and operated by the farm family. During the time period (1998 – 2002) when a large part of the data used in this study was collected, the average herd size was about 30 cows. Due to a major structural change that is taking place at the moment, the average herd had grown to 46 cows in 2005 (Statistics Sweden, 2006), a figure that suggests that the farms grow quickly. However, many dairy farms also exit dairy production. Farm production at a Swedish dairy farm is often diversified, where the farms have major beef production and also cash crop production. Most farmers use tied milking system, although milking parlours and automatic milking systems are getting more common. The data set used in this research project is stratified according to farm size and geographic location.

To conduct the research synthesized in this chapter with the research method outlined in the previous section, information about how much inputs each farm use, how much outputs they produce, the prices of the inputs and outputs, and data detailing the study objects presented in Table 1 (the factors explaining efficiency) were needed. Some of these data were available in existing data sources, whereas the remaining data had to be collected through a mail questionnaire.

5.1. Inputs, Outputs and Prices

To construct the input and output variables needed to estimate the efficiency scores, farm level accounting data available from Statistics Sweden were used. For some variables, prices could be calculated from the accounting data, but when that was not possible, prices were obtained from a database consisting of gross margin budgets for different agricultural production lines and regions in Sweden (www.agriwise.org, 2005).

Statistics Sweden collects numerous data from farms, with the main purpose to be the basis of the Farm Accounting Data Network (FADN). The data consist of the balance sheet and income statement of each farm in the sample, as well as of some additional data such as time worked at the farm and harvest. Statistic Sweden uses a rotating panel to construct the dataset. This means that not all farms are represented each year. Data collected in 1998 through 2002 were used. As an attempt to compensate for stochastic variation to which DEA is sensitive, each farm were represented by its own yearly average of inputs outputs and prices during the years it participated in the dataset. A little more than 500 farms were included in the data set, which corresponded to about 4% of the population of milk producing farms in Sweden at the time the data was collected.

Inputs were aggregated into the six major ones for a dairy farm: fodder, labour, capital, energy, seed and fertilizer. Five final outputs were considered: milk, livestock, forage, crop and "other". "Other" corresponds to all other outputs at the farm, but consists mostly of revenues from governmental supports.

5.2. Measures of Driving and Restraining Forces

Some of the data needed to construct measures of driving and restraining forces for successful dairy farm performance were available in the dataset from Statistics Sweden. For example, measures of farm size and the age of the farmer were available in that dataset. Information on management's critical success factors (MCSFs) came from a dairy cow recording scheme conducted by the Swedish Dairy Association. Participation in this recording scheme is voluntary to the dairy farmers, but most of the farmers do participate. On a monthly basis the recording scheme reports information of cow level performance, in terms of, for example, milk yield, milk quality, diseases and fertility to the farmer.

The remaining data needed were collected through a mail questionnaire. The development of the questionnaire started with literature reviews to find possible determinants of efficiency and a focus group meeting where a group of dairy farmers were interviewed about how they believed farms could become more efficient. During the development time the questionnaire was tested in three subsequent focus group interviews where dairy farmers were asked to fill in the questionnaire and to explain how they interpreted the questions. The questionnaire was sent to the farmers who participated in the data collection by Statistics Sweden, referred to above in the beginning of February 2005. After two reminders, the response rate was 65%, however, there were some problems with questionnaires only partly filled out which caused missing variables.

5.3. Discussion about the Quality of the Data

Naturally, the quality and validity of all kinds of studies depend heavily on the quality of the data used, and the research synthesized in this chapter is no exception. Keeping the quality of the data in mind when interpreting the results is always important. As in other studies in social sciences much of the data collected for this research is based on opinions of humans, which of course may cause measurement errors. This is especially important for the variables that originate from information collected through the questionnaire. For example, the mode of the person who answers a questionnaire may affect the answers. Tiredness, amount of time devoted to the questionnaire, motivation, expectations and other factors may influence the answers. Furthermore, it is possible that questions and answer models are interpreted differently by different individuals. Especially questions where the respondent is asked to rate something may be answered differently by different individuals, where "good" and "bad" may be interpreted differently by different individuals. For these reasons, the estimated coefficients of that kind of variables should be interpreted with care, only as directions. To minimize measurement errors in the questionnaire, actions such as focus groups interviews were undertaken to improve the transparency of the questions and the answer models.

The data used to calculate the efficiency scores were, as mentioned above, accounting data from Statistics Sweden and price data from Agriwise (www.agriwise.org, 2005). Accounting data can be subject to measurement errors if there is inconsistency in the classifications of the transactions into accounts at the farms and because of the subjectivity of, for instance, the valuations of depreciations. However, because Statistics Sweden uses an accountancy agent to collect the data, there are no particular reasons to believe that there are extensive errors of the above-mentioned kind. Accounting data are, however, based on historical values, which may cause problems especially with the valuation of capital, in this research measured as the value of buildings, other fixed equipment and production rights. The real value may therefore differ from the booked value. However, possible actions to correct for this, such as valuing all the capital at all farms would be very expensive and time consuming so these were not considered. When individual price data are not available in the dataset from Statistics Sweden they were collected from Agriwise (www.agriwise.org, 2005). This means that the same price is assumed for all farms in a particular geographic region. Therefore, differences in prices due to, for instance ability to negotiate are not included in the data. Furthermore, because quantities of inputs and outputs were for some variables calculated by dividing expenses and income with the relevant price, this means that at large farms, which may have an advantage in negotiating prices, the use of inputs may have been underestimated, and the amount of outputs overestimated. This in turn, may cause efficiency to be overestimated at these farms.

Because the data panel from Statistics Sweden ended in 2002, pooling it with the data from the questionnaire collected in the beginning of 2005 obviously implies that data from different years are pooled. If some farmers have made large changes to their dairy farms, such as investing in new barns or milking systems after 2002, this is reflected in the questionnaire but not in the accounting data from Statistics Sweden and this can influence the results. If flaws like this have affected the data, they are more likely to affect variables describing investments in fixed assets at the farm than values and management routines, which are more likely to be stable over time. However, because the coefficients of the variables describing investments in fixed assets are logical in the estimations, these possible flaws are not thought to be any problem.

Out of the data used in this research project, the data from the Swedish Dairy Association are the data that are the least likely to be subject to measurement errors. The same product is evaluated at each farm with the same type of instruction for milk collection procedures and analysis devices in each case. Based on experience gained from the work with this research project, I think that the optimal way to collect data for such a study would be by telephone interviews. Ideally, a battery of questions would be constructed and sent to the farmers well in advance of the telephone interview, so that they can prepare answers to the questions. With this approach, difficult questions can then be explained directly, and all questions are likely to be answered. However, this approach would be very time consuming and thus expensive both for the researcher and the farmers. Further, the fact that people perceive words such as "good" or "bad" differently will still remain. As a consequence, given the time and budget constraints that everyone doing research is subjected to, the dataset used in this research project is satisfactory.

6. A Synthesis of the Empirical Findings

A convenient way to synthesize the different studies that constitute the research project presented in this chapter is to start at a broad level and subsequently refine and detail the analysis. The analysis was started by trying to broadly understand what the main reason is for low efficiency in dairy farms and to evaluate usefulness of commonly used management's critical success factors (MCFSs). This was done in Hansson, 2007a, hereafter called the MCFSs study. Further, the effects of farm size were analyzed in the beginning of the study, in Hansson, 2007b, hereafter referred to as the size study. In Hansson (2007c), the effects of strategy drivers and restraints on farm efficiency were analyzed. This study is hereafter referred to as the strategy study. In Hansson and Öhlmér (2008) the effect of differences in operational managerial practices, which can be adjusted in the daily farm work was studied. This study is henceforth referred to as the operational managerial practices study. Finally, in Hansson (2008) the effect of differences in managerial capacity of the farmer were studied. This study is referred to as the managerial capacity study in the remaining parts of this chapter.

Table 3 summarizes the studies with respect to research questions as well as to study object and categories of analyzed factors, in light of Table 1.

Table 3. Summary of papers in the research project. The list is organized according to research questions as well as to study object and categories of analyzed factors

	Research questions	Study object and categories of analyzed factors
The MCSFs study (Hansson 2007a).	· What part of the profitability process is the more problematic? · Are there any links between management's critical success factors and farm level efficiency?	*The overall structure of efficiency* · Comparison between long-run input and output efficiency scores. · Analysis of how well some common management's critical success factors from a dairy cow recording scheme work in predicting good economic performance.
The size study (Hansson 2007b).	· Are larger farms more efficient?	*The overall structure of efficiency* · How does farm size, in terms of the size of the dairy production, the whole-farm production and the number of hectares affect efficiency?
The strategy study (Hansson 2007c).	· What are the effects of factors decided in the long-run strategic decision making?	*The external-operational, internal and micro-social environments* · Geographic location. · The outcome of long-run decisions about resource allocation and use. · The outcome of long-run decisions about fixed costs. · The social situation of the farm
The operational managerial practices study (Hansson and Öhlmér, 2008).	· What are the effects of factors adjustable in the short-run operational managerial practices?	*Short-run operational managerial practices* · Animal health practices · Breeding practices · Feeding practices
The managerial capacity study (Hansson 2008).	· What are the effects of the managerial capacity of the farmer?	*The farmer managerial capacity* · Values and attitudes. · Psychological aspects. · Experience. · Education. · Search for information for decision making. · Planning, forecasting and evaluating consequences for decision making. · Bearing responsibility in decision making.

As indicated in Table 4, the first two parts of the research project analyzed the overall structure of efficiency among the dairy farms in Sweden. The third part analyzed how factors that are decided in the long-run strategic decision making affect farm efficiency. Knowledge of this is important especially when new farms are started or when old ones undergo major changes. The fourth part analyzed how management practices adjustable in the daily work affect efficiency. The fifth part, finally, analyzed how farmer managerial capacity can contribute to improved farm performance. Following this introductory text, this section continues by summarizing and discussing the main findings in each of the five studies. However, before that is done, descriptive statistics of the efficiency scores are presented.

6.1. Summary Statistics of the Efficiency Scores

Throughout the research project, the analyses were based on efficiency scores estimated from a sample of 507 milk producing farms. Inclusion in this sample was based on the criteria that at least 50% of the farm total income from milk, livestock, crops and forage should originate from milk. However, in the size study, it was motivated to make the inclusion in the analyzed sample less generous. Therefore, inclusion in the size study was based both on a size criteria and on a specialization criteria. Table 4 shows summary statistics of all nine major efficiency scores, based on the sample of 507 milk producing farms.

It is important to stress that efficiency is a relative concept, based on the studied sample, where the efficiency of each farm is determined in light of its performance relative to the other farms in the sample. Factors such as sample size and the variable specification can influence the results (see, e.g., Coelli et al. 2005 for a discussion on this). Therefore it is not meaningful to compare the descriptive statistics based on this research project to those of other research projects. Such comparisons are no firm basis to argue that, e.g., farms in one country are more efficient than farms in an other country. Such a comparison could only be meaningful if observations from different countries were included in the same empirical estimation.

Table 4. Descriptive statistics of the efficiency scores. The statistics are based on a sample of 507 milk producing farms.

Efficiency type	Mean	Minimum	Standard deviation	Skewness
Input-oriented efficiency, long run				
Economic efficiency	0.645	0.119	0.165	0.366
Technical efficiency	0.865	0.410	0.148	- 0.756
Allocative efficiency	0.752	0.119	0.161	- 0.528
Input-oriented efficiency, short run				
Economic efficiency	0.616	0.118	0.242	0.212
Technical efficiency	0.889	0.282	0.165	- 1.213
Allocative efficiency	0.692	0.118	0.226	- 0.218
Output-oriented efficiency				
Economic efficiency	0.745	0.240	0.180	- 0.106
Technical efficiency	0.854	0.276	0.164	- 0.855
Allocative efficiency	0.873	0.294	0.117	- 1.582

The highest possible efficiency score is 1.000.

However, having noticed this weakness of efficiency studies, it is indeed interesting to compare the efficiency scores presented in Table 5 with each other. For instance, the allocative efficiency scores are consistently lower than the technical efficiency scores in both the long- and short-run input-oriented cases. This suggests that the main reason for low economic input efficiency is low allocative efficiency, i.e., a cost-minimization problem. In the output-oriented case, allocative efficiency is the higher measure, suggesting that in this case technical efficiency is the main reason for low overall economic efficiency. Furthermore, it is also interesting to note that the main efficiency problems seem to lie within the short-run cost-minimization, where the average economic efficiency score suggests a possibility to decrease costs by almost 40% if all farms were as efficient as the most efficient ones. The skewness of the efficiency scores suggests that technical and allocative efficiency scores are typically skewed towards full efficiency, whereas the economic efficiency scores are typically skewed towards lower efficiency levels.

6.2. The MCSFs Study

In the MCSFs study (Hansson 2007a) the aim was to study the relationship between economic input-oriented, economic output-oriented and profit, i.e., both economic input and output-oriented efficiency, and some management's critical success factors (MCSFs) that are commonly used at dairy farms. By doing this, the usefulness of the MCSFs in indicating good economic performance could be evaluated.

The MCSFs considered in MCSFs study were milk yield per cow, protein content in the milk, average herd fertility, mastitis ratio and involuntary culls ratio. These MCSFs are particularly interesting to study because they are all reported on a monthly basis to all dairy farmers who participate in the dairy cow recording scheme conducted by the Swedish Dairy Association. Thus, the considered MCSFs can be used regularly at the farms to indicate the development of the farm performance, and are therefore likely to influence the behaviour of the farmers.

In the MCSFs study, all farms that scored one on the economic input-oriented efficiency score were defined as cost efficient, all farms that scored one on the economic output-oriented efficiency score were defined as revenue efficient and finally, all farms that scored one on both the cost and revenue efficiency scores were defined as profit efficient. These scores were defined as success in logistic regressions, where the dependent variables were the MCSFs defined above. The results showed that only two of the considered MCSFs worked as significant predictors of good economic performance. Milk yield per cow predicted being cost efficient and being profit efficient. The mastitis ratio was a significant predictor of not being revenue efficient. None of the MCSFs protein content, fertility or the involuntary culls ratio were significant predictors of any definition of good farm performance. Therefore, the results of the MCSFs study suggests that actions at the farm to improve the protein content, fertility or the involuntary culls ratio to the levels of the most efficient farms may not be worthwhile. Reasons may be that all farms already perform relatively well on these scores, regardless of farm efficiency level.

6.3. The Size Study

The size study (Hansson 2007b) aimed to investigate how economic, technical and allocative input-oriented efficiencies were affected by differences in farm size. Also, the paper aimed to investigate the scale efficiencies and effects to further study the potential improvements in efficiency due to larger farms. Given the structural change in Swedish dairy production, outlined in the introductory section of this chapter, an understanding of the effect on efficiency of farm size is of great importance. In the size study, three measures of farm size were analyzed: the size of the dairy production, the size of the whole-farm production and the physical size of the farm in terms of the number of hectares. Both linear and squared effects of these size measures were considered. A sub-sample of specialized dairy farms was studied separately as well as together with the full sample of milk producing farms.

The size study showed that the connections between farm level efficiency and size are involved and that farm size does not influence all efficiency scores in only one direction. For instance, the linear effects of the size of the dairy production were significantly negative for technical efficiency in both groups of farms and also for economic efficiency in the full sample. None of the linear size effects influenced allocative efficiency.

When the squared effects were considered, technical efficiency was first decreasing with the size of the dairy production and the number of hectares and then increasing. Allocative efficiency, on the other hand, was first increasing with the size of the dairy production and the number of hectares at the farm, and then the allocative efficiency was decreasing. Significant effects of the squared measure of the whole-farm production suggested that technical efficiency first increased and then decreased with farm size, whereas the effect was reversed for allocative efficiency. The size study thus showed that technical and allocative efficiency scores are typically affected by farm in opposite ways size. Compared to previous literature, the results were different from those of Bravo-Ureta and Rieger (1991), but similar to the results reported by Helfand and Levine (2004).

The size study reported high average scale efficiency in both the full sample and in the sub-sample, 95.3% and 96% respectively. These results suggests that, on average, the farms operate close to their optimal scales and would not benefit much from moving closer to their optimal scales. The study also showed that in both groups several farms were operating under increasing returns to scale, which means that these farms could increase their efficiency by expanding their farms. However, also these farms have high average levels of scale efficiency, 91.8% based on the full sample and 91.7% based on the specialized sub-sample, implying that the gains from increasing farm size are small.

6.4. The Strategy Study

In the strategy study (Hansson 2007c) the aim was to study the impact of factors that work as driving or restraining forces on the farm strategy, and therefore on farm efficiency. Building on work by Lee et al. (1999), Gasson et al. (1988) and Harling and Quail (1990) a model was developed, consisting of three levels of environments: external-operational, internal and micro-social. All environments contain potential driving and restraining forces for farm strategy and therefore also for farm performance. The external-operational environment is difficult to influence by the individual farmer and therefore treated as given.

The internal and micro-social environments are to some extent controllable by the farmer, at least in the long run and therefore interesting to consider in the strategic management.

In the strategy study, all nine efficiency scores (see Table 2) were calculated and the effects of factors that determine the above-mentioned environments, and therefore also determine farm strategy, were analyzed. The results showed that several strategy factors were important for farm performance. For example, several geographic locations (determinants of the external-internal environment) were driving forces for economic and technical efficiency. Therefore, the results suggest that some parts of Sweden (especially the southern ones) are better equipped for dairy production than the north parts. As elaborated in the strategy study, differences in soil and climate may be reasons, but also potential differences in business culture.

When the determinants of the internal farm environment was considered, the strategy study found that size of fields, distance to fields, bunker silo and tower silo were driving forces for farm performance, while quality of forage machinery and a high focus on dairy production were found to restrain farm performance. As suggested in the strategy study, the fact that high quality of forage machinery restrains farm performance suggests that the farmers have over-invested in forage machinery leading to too much capital tied in the forage machinery. The study also suggests that high investments in certain production lines may cause the farmer to be less willing to switch to other production lines even if they are more profitable at the time. Interestingly, the strategy study also found that diversified farms are more efficient than specialized ones. This result was supported by findings by Hadely (2006) and Brümmer (2001). Reasons elaborated in the strategy study can be that diversified farms are likely to have other fully developed production lines that they can expand when the market conditions are more favourable for these and thus resources are allocated where they are the more profitable at the time. Another reason that was stressed in the strategy study suggested that that milk production may be more difficult or less profitable compared with other agricultural production lines.

Within the micro-social environment the strategy study showed that co-farmers were restraints on farm performance, which may indicate a hidden unemployment problem at the farm. Furthermore, the results in the strategy study showed that discussing dairy production with someone improved the skills to minimize costs. This result therefore suggested that there is a learning process in the micro-social environment where farmers get ideas from each other how inputs should optimally be combined.

6.5. The Operational Managerial Practices Study

In the operational managerial practices study (Hansson and Öhlmér 2008) the aim was to study how operational managerial practices influence farm efficiency. Operational managerial practices were defined as practices that can, if not be totally changed, at least improved in the daily farm work. Operational managerial practices were defined as animal health factors, breeding practices and feeding practices. Among the animal health factors, age of the heifer at her first calving, length of calving interval and length of dry period were considered. Among the breeding practices, differences in the breeds and whether or not the breed percentage equalled the replacement percentage were considered. Among the feeding practices, finally, practices such as analysis of forage and fodder grain, calculation of feed ration and the mix of forage were considered. Focus was only on the input-oriented efficiency scores, because operational managerial practices should be of importance especially for the cost-minimization problem at the farm rather than for the revenue-maximization.

The results of the operational managerial practices study showed that managerial practices adjustable in the daily farm work typically do not affect farm efficiency to any larger extent. Only four out of the nine considered managerial practices influenced efficiency significantly and in general only at the 10%-level. Typically, the significant managerial practices were significant only on single occasions (e.g. none of the significant managerial practices influenced all efficiency scores significantly). Most significant cases were found among the feeding practices.

Interestingly, none of the animal health factors significantly affected the efficiency scores in the operational managerial practices study. These results are to the contrary of experimental studies reported by Sorensen and Ostergaard (2003) and Bertilsson et al. (1997), where animal health factors similar to those considered in the operational managerial practices study were studied in light of economic results. A reason for contradicting results that we discussed in the operational managerial practices study is the empirical setting of our research, where the best practice farms were defined by the most efficient empirical frontier. This may differ from results obtained in experimental settings. A further reason discussed in the paper is the whole-farm approach applied in our study, compared to the enterprise approach applied in other studies. Efforts to optimize one enterprise may lead to other enterprises being neglected and therefore the whole-farm efficiency may be unaffected.

Obviously the results reported in the operational managerial practices study give the impression that at least some managerial practices, and especially the animal health practices, do not matter for farm efficiency. Similar impression was also obtained in the MCSFs study, as indicated above. Is it really so? No, that would be an over-simplified explanation. A more realistic explanation elaborated both in the operational managerial practices study and in the MCSFs study is that all farms apply similar managerial practices, and especially similar animal health practices, regardless of the individual level of efficiency. Almost all dairy farms in Sweden participate in a dairy farm recording scheme conducted by the Swedish Dairy Association, which give similar managerial advice to all farms. Therefore, all farms use similar managerial practices. The less efficient farms can then typically not learn how to become more efficient by adapting to the operational managerial practices used by the more efficient farms, because the differences between the farms are almost non-existing.

6.6. The Managerial Capacity Study

In the managerial capacity study (Hansson, 2008) the aim was to investigate the impact of managerial capacity on farm efficiency. A model was developed where efficiency was modelled as dependent on personal characteristics of the farmer and on management systems aspects, which serves to facilitate especially analytical decision making. Because personal aspects of the farmer contributes to a readiness to act in a certain way, based on the tacit system as described by Hogarth (2001), and because previous empirical findings (Öhlmér et al. 1998; Öhlmér and Lönnstedt 2004) found that farmers generally prefer decision making based on tacit, intuitive systems, the link between personal characteristics and efficiency was expected to be stronger than the link between management systems and efficiency.

The managerial capacity study was conducted in three independent steps. In the first step economic, technical and allocative long- and short-run input efficiencies were calculated with DEA (Charnes et al. 1978) as well as economic, technical and allocative output efficiencies. In a second step the direct effect of the managerial capacity aspects were estimated using tobit regressions. In the third step, logistic regressions were used to estimate the effect of personal aspects on the significant management systems aspects.

The results of the managerial capacity study showed that several of the managerial capacity aspects are important determinants of farm efficiency. Interestingly, this holds especially for the input-oriented efficiency scores. The study found that personal aspects such as a positive profitability attitude towards dairy farming today, internal locus of control, an education in agriculture, participation in study circles and intrinsic values affected farm level efficiency in a positive way. Further, the managerial capacity study found that values of the farmer influenced short-run input-oriented efficiency scores to a greater extent than the long-run input-oriented efficiency scores and the output-oriented efficiency scores. A possible explanation for this, as elaborated in the paper, is that in the short-run intuitive decision making is more important because action alternatives are well known. In the intuitive thinking process, values are important (Klein et al. 2005).

The managerial capacity study also showed that positive expectations about future profitability in dairy farming affected efficiency negatively. A reason emphasized in the study is that farmers who are positive about the future may have made investments to which they have not yet adjusted. Further, expressive goals, age and profitability perception also affected efficiency negatively. Based on how profitability perception was measured, this implies that believing that the farm is better than it is leads to lower efficiency.

The managerial capacity study showed that personal aspects of the farmer were more important for efficient farming compared to management systems aspects, with more significant variables affecting farm efficiency. This outcome was expected because previous literature (Öhlmer at al. 1998, Öhlmér and Lönnstedt, 2004) has found that farmers generally prefer intuitive, tacit decision-making. Personal aspects contribute to a persons general experience and therefore facilitate the intuitive, tacit thinking processes (Hogarth, 2001), whereas management systems support analytical decision making. In fact none of the considered management systems aspects influenced the output-oriented efficiency scores significantly. A further reason as suggested in the study is that it is difficult to influence output prices and the output mix of a Swedish dairy farmer. Among the significant management systems aspects, the study found that farmers who find arm advisors and other farmers or colleagues important information sources in the decision making process, generally have higher input-oriented efficiency. Further, the managerial capacity study found that checking the accounting and paying attention to collected information had positive effects on efficiency.

Instrumental and expressive goals, as well as participation in study circles, were aspects found to influence the management systems aspects that influenced efficiency positively. Based on the results, the managerial capacity study suggested a concrete way of supporting dairy farms to become more profitable and that is by organizing combined educational and discussion clubs where the farmers get to learn from each other as well as from professional dairy farm advisors.

6.7. Summary of the Main Empirical Findings

Table 5 summarizes the main conclusions of the five different studies in light of Table 1. The table shows that there are several reasons for inefficient production in Swedish dairy farms, and that reasons for inefficient production can be understood both on a structural level as well as on the farm and farmer characteristics level.

Table 5. Summary of the five different studies in light of Table 1

Theme and studies	Conclusions
The structure of inefficiency The MCSFs study (Hansson, 2007a). The size study (Hansson, 2007b).	· The input, or cost, perspective offers more opportunities for improvements in efficiency. Especially the allocative efficiency can be improved. · The links between MCSFs and economic performance according to the efficiency scores are not necessarily strong. Only two MCSFs out of five were significant predictors of efficiency. · The relationships between farm size and farm level efficiency are involved and the results show that the technical and allocative efficiency scores are typically affected in opposite ways by farm size.
The farm itself The strategy study (Hansson, 2007c).	· External-operational environment in terms of geographic location can drive efficiency. · In the internal environment, size of fields, distance to fields, bunker silo and tower silo drive efficiency. High quality of forage machinery and high focus on dairy production restrain efficiency. · In the micro-social environment, a discussion partner drives efficiency, while co-farmers restrain efficiency.
Operational managerial practices The operational managerial practices study (Hansson and Öhlmér, 2008).	· Operational managerial practices have only limited possibilities to improve farm efficiency, if the inefficient dairy farmers adjust to the operational managerial practices of the best farmers.
The farmer managerial capacity The managerial capacity study (Hansson, 2008).	· Of the personal aspects of the farmer, positive ideas of profitability in dairy farming today, internal locus of control, education in agriculture, participation in study circles and intrinsic values drive efficiency. A positive expectation about profitability in future dairy farming, expressive goals and age restrain efficiency. · Management systems aspects found to drive efficiency were checking accounting, paying attention to collected information and information sources such as farm advisors, other farmers and colleagues.

7. Conclusion

The aim of this chapter was to synthesize, in light of previous literature, the main findings of five separate empirical studies, all with the common aim of analyzing what drives and restrains efficient production in dairy farms, to conclude on concrete actions to strengthen the farms and help them become more efficient. From the synthesis of the literature review above (Section 1.2.3) four themes emerged as fruitful bases to analyze why some farms are efficient and others are not: the overall structure of efficiency, the farm itself, operational managerial practices and the farmer managerial capacity. The research project synthesized in this chapter was centred on these themes.

Most significant differences between the efficient and the inefficient farms were found in the strategy study and in the managerial capacity study. However, analysis of the overall structure of the efficiency (the MCSFs study and the size study) also gave important insights. These results showed that especially the ability to minimize costs, i.e., to combine inputs in the cheapest way, needs to be improved. Further, the MCFSs study showed a weak connection between MCSFs and economic performance. This result is alarming because it shows that actions at the farm to strengthen the MCSFs do not necessarily lead to better economic performance. The results also indicate that the best farms do not have better scores on the MCSFs compared with the less efficient farms. Reasons for this may be that all farms participating in the dairy cow recording scheme receive similar managerial advice, similarly to the suggestions reported to explain insignificant results of the factors considered in the operational managerial practices study. The analysis of the overall structure of efficiency also showed that farm size affects efficiency in different ways depending on what type of efficiency is considered. For example, farm size typically affected technical and allocative efficiencies in opposite ways. These results have important implications for farmers who plan to expand their farms because the expansion will not necessarily lead to a more efficient farm.

The strategy study showed that aspects decided in the long-run strategic management, such as the location of the farm, the farm layout (size of fields and distance to fields), barn type, equipment for forage production and degree of focus on dairy production lay a long-term basis for the farm efficiency and explain differences in efficiency. The strategy study also showed the importance of the micro-social environment. For instance, discussing the dairy production with someone on a regular basis leads to more efficient farms.

When it comes to the farmer managerial capacity (the managerial capacity study), the results show that the more successful farmers have more intrinsic values, have a high degree of internal locus of control, participate in study circles and have an education in agriculture. Interestingly, the results also showed that farmers who believe that they are better than they really are, have lower levels of efficiency. Further, farmers with a positive idea about their own profitability are also more efficient. The results also show that more successful farmers use management systems such as seeking information from farm advisors and colleagues, checking the accounting often and paying more attention to gathered information compared to unsuccessful farmers.

Based on these results synthesized in this chapter, how can the farmers be supported to become more efficient? One suggestion that emerges from the research is to focus on developments of the management skills of the farmers. For instance, the farmers need to become better at minimizing costs. This involves developing skills of the farmers such as valuing inputs that are difficult to value (e.g., own labour time and equity financed capital). The results further show that suitable forums for this are forums where the farmers get to learn both from other colleagues as well as from farm advisors. Consequently, the farms can be supported to become more efficient and thus more profitable by organizing combined educational and discussion clubs. The farmers would then not only be able to learn from each other but also inspire each other. The results further suggest that focus needs to be shifted away from the biological concerns, towards the behaviour and strategic management of the farmers.

7.1. Opportunities for Further Research

Upon synthesizing this four-year research project aiming at analyzing what drives and what restrains efficient production at dairy farms, two lines of research, in particular, emerged as fruitful continuations of the research. Before ending this chapter, these two lines are briefly summarized.

7.1.1. Suggested Research Line 1: The Impact of External Factors

To a large extent, an internal perspective to analyzing why some farmers succeed in their farming while others do not has been applied in this research project. The reasoning has been that internal factors are adjustable by the individual farmer, at least in the long-run strategic management. However, running a successful farm business, or any kind of business, is dependent on the infrastructure, in the widest sense of the word. Pfeffer and Salancik (1978) offer, as mentioned in the literature review, a resource dependence perspective to describe how firms act and develop. This means that actions of firms are understood by analyzing the context in which they operate. Therefore, within this framework, the development of firms is determined by their interdependence with other firms and individuals in the environment. Henrekson (2001) discussed the institutional prerequisites in Sweden in light of entrepreneurship, and concluded that unfavourable prerequisites in Sweden have constrained the founding of small-scale firms. A study that illustrates how the environment, in terms of both other firms and individuals as well as institutional prerequisites, affect successful farming would shed further light on what drives and restrains efficient farming. Furthermore, it would be interesting to analyze how efficient and thus successful farms act on the environment in which they operate, compared to the less successful farms. Such a study could involve how the farmer acts on a broad range of external interested parties such as creditors, the dairy plant processor, political decisions, consumers, other farmers and farm advisors.

7.1.2. Suggested Research Line 2: The Information Needs

As was stressed in Klein et al. (2005), interpreting information from the environment is a foundation to be able to affect one's situation. In business management, information has a substantial impact, not only because it shows the firm results and fulfilments of business goals in other respects, but also because it contributes to the detection of problems, opportunities and threats. When analyzing statistics about size and the number of dairy farms in Sweden a reality is that the farms are becoming larger and larger. Traditionally, dairy farms in Sweden have been one-man or one-woman businesses, or family businesses. However, with the present development towards larger farms, this situation is likely to change, and with that more people are likely to work at the farm. For instance, the increased need of capital will trigger growth in the number of business owners. With these developments, management will become even more important. Questions such as when to adapt to new technology, how to finance the business and how to manage employees are likely to get more attention in the future. With these developments the ability to act on relevant information both at the farm and in the farm environment becomes more critical. However, given that the usage of the deliberate system is costly (Hogart, 2001) the value of information is limited by the design on information. For instance, Öhlmér and Lönnstedt (2004) concluded that traditional accounting statements are not as useful as verbal interpretations of them, even to farmers with an analytical approach to interpreting information. Still, information intended for farm managers, such as accounting reports and reports from the dairy cow recording scheme, is designed to fit analytical thinking processes. The actual information needs at dairy farms have only been studied in a few studies (Huirne et al. 1998; Asseldonk et al. 1999; Lunneryd 2003), but a detailed study that links information use and interpretation of information to the questions of economic, technical and allocative efficiency is not found in previous literature. Given this, a suggestion for further research is to analyze the information needs at dairy farms as well as to analyze how information use and interpretation of information differs between the best practice farms and the remaining ones.

Acknowledgements

The research synthesized in this chapter originates from a project which partly constituted my PhD thesis. The research project has benefited very much from advice from Prof. Bo Öhlmér and Prof. Hans Andersson, both at the Swedish University of Agricultural Sciences. Financial support was obtained from the Swedish Farmers' Foundation of Agricultural Research, which is gratefully acknowledged.

References

Agriwise: *http://agriwise.org/* visited January 10-11 2005, October 2005. Department of Economics, Swedish University of Agricultural Sciences.

Afonso, A. and St. Aubyn, M. (2006). Cross-country efficiency of secondary education provision: A semi-parametric analysis with non-discretionary inputs. *Economic Modelling, 23,* 476-491.

Aigner, D., Lovell, C. A. X and Schmidt, P. (1977). Formulation and Estimation of Stochastic Frontier Production Function Models. *Journal of Econometrics, 6,* 21-37.

Alvarez, A. and Arias, C. (2004). Technical efficiency and farm size: a conditional analysis. *Agricultural Economics, 30,* 241-250.

van Asseldonk, M. A. P. M., Huirne, R. B. M., Dijkhuizen, A. A., Beulens, A. J. M. and Udenk ten Cate, A. J. (1999). Information needs and information technology on dairy farms. *Computers and Electronics in Agriculture, 22,* 97-107.

Barney, J. (1991). Firm resources and sustained competitive advantage. *Journal of Management, 17,* 99-120.

Barney, J., Wright, M. and Ketchen, Jr. D. J. (2001). The resource-based view of the firm: Ten years after 1991. *Journal of Management, 27,* 625-641.

Bertilsson, J., Berglund, B., Ratnayake, G., Svennersten-Sjaunja, K and Wiktorsson, H. (1997). Optimising lactation cycles for the high yielding dairy cow. A European perspective. *Livestock Production Science, 50,* 5-13.

Boehlje, M. D. and Eidman, V. R. (1984). *Farm Management.* New York: John Wiley and Sons.

Bravo-Ureta, B. E. and Rieger, L. (1991). Dairy farm efficiency measurement using stochastic frontiers and neoclassical duality. *American Journal of Agricultural Economics*, *73,* 421-428.

Brümmer, B. (2001) Estimating confidence intervals for technical efficiency: the case of private farms in Slovenia. *European Review of Agricultural Economics, 28,* 285-306.

Charnes, A., Cooper, W. W., and Rhodes, E. (1978). Measuring the Efficiency of Decision Making Units. *European Journal of Operational Research, 2,* 429-444.

Coelli, T. (1995). Recent developments in frontier modeling and efficiency measurement. *Australian Journal of Agricultural Economics, 39,* 219-245.

Coelli, T. and Perelman, S. (1999) A comparison of parametric and non-parametric distance functions: With application to European railways. *European Journal of Operational Research*, *117,* 326-339.

Coelli, T. Rahman, S. and Thirtle, C. (2002). Technical, allocative, cost and scale efficiencies in Bangladesh rice cultivation: a non-parametric appraoch. *Journal of Agricultural Economics, 53,* 607-626.

Coelli, T., Rao, P. D. S., O'Donnell, C. J. and Battese, G. E. (2005). *An introduction to efficiency and productivity analysis*. New York: Springer Science+Business Media, LLC.

Cullinane, K., Wang, T-F., Song, D-W. and Ji, P. (2006). The technical efficiency of container ports: Comparing data envelopment analysis and stochastic frontier analysis. *Transportation Research Part A, 40,* 354-374.

Cyert, R. and March, J. (1963). *A Behavioral Theory of the Firm*. New Jersey: Prentice-Hall Inc.

Daft, R. L. 2003. *Management.* Mason: Thomson Learning.

Farrell, M. J. (1957). The measurement of productive efficiency. *Journal of the Royal Statistical Society. Ser. A, 120,* 253-281.

Gasson, G. (1973). Goals and values of farmers. *Journal of Agricultural Economics, 24,* 521-542.

Gasson, R., Crow, G., Errington, A., Hutson, J., Marsden, T. and Winter, D. M. (1988). The farm as a family business: a review. *Journal of Agricultural Economics, 39,* 1-41.

Hadely, D. (2006). Patterns in technical efficiency and technical change at the farm-level in England and wales, 1982-2002. *Journal of Agricultural Economics, 57,* 81-100.

Hansson, H. (2008). How can farmer managerial capacity contribute to improved farm performance? A study of dairy farms in Sweden. *Food Economics - Acta Agricult Scand C, 5, 44-61.*

Hansson, H. and Öhlmér, B. (2008). The effect of operational managerial practices on economic, technical and allocative efficiency at Swedish dairy farms. *Livestock Science, 118, 34-43.*

Hansson, H. (2007a). The links between management's critical success factors and farm level economic performance on dairy farms in Sweden. *Food Economics - Acta Agricult Scand C, 4,* 77-88.

Hansson, H. (2007b). Are larger farms more efficient? A study of the relationships between farm level efficiency and size in Swedish dairy farms. Unpublished draft. *Department of economics, Swedish University of Agricultural Sciences.*

Hansson, H. (2007c). Strategy factors as drivers and restraints on dairy farm performance: Evidence from Sweden. *Agricultural Systems, 94*, 726-737.

Harlig, K., F. (1992). A test of the applicability of strategic management to farm management. *Canadian Journal of Agricultural Economics, 40,* 129-139.

Harling, K. F. and Quail, P. (1990). Exploring a general management approach to farm management. *Agribusiness, 6,* 425-441.

Hayashi, F. (2000). *Econometrics.* New Jersey: Princeton University Press.

Helfand, S., M. and Levine, E. S. (2004). Farm size and the determinants of productive efficiency in the Brazilian Center-West. *Agricultural Economics, 31,* 241-249.

Henrekson, M. (2001). Institutionella förutsättningar för entreprenörskap och företagstillväxt. In: Davidsson, P., Delmar, F. and Wiklund, J. (eds) *Tillväxtföretagen i Sverige.* (pp 38-82) Stockholm: SNS Förlag.

Heshmati, A. and Kumbhakar, S. C. (1994). Farm Heterogeneity and Technical Efficiency: Some Results from Swedish Dairy Farms. *The Journal of Productivity Analysis, 5,* 45-61.

Hogarth, R. M. (2001). *Educating Intuition.* Chicago: The University of Chicago Press.

Huirne, R. B. M., Harsh, S. B. and Dijkhuizen, A. A. (1997). Critical success factors and information needs on dairy farms: the farmer's opinion. *Livestock Production Science, 48,* 229-238.

Iráizoz, B., Rapún, M. and Zabaleta, I. (2003). Assessing the technical efficiency of horticultural production in Navarra, Spain. *Agricultural Systems, 78,* 387-403.

Kennedy, P. (1998). *A guide to econometrics.* Oxford: Blackwell publishers inc.

Klein, G., Pliske, R., Crandall, B. and Woods, D. D. (2005). Problem detection. *Cognition Technology and Work, 7,* 14-28.

de Koeijer, T. J., Wossink, G. A. A., Smit, A. B, Janssens, S. R. M., Renkema, J. A. and Struik, P. C. (2003). Assessment of the quality of farmers' environmental management and its effects on resource use efficiency: a Dutch case study. *Agricultural Systems, 78,* 85-103.

Latruffe, L., Balcombe, K., Davidova, S. and Zawalinska, K. (2005). Technical and scale efficiency of crop and livestock farms in Poland: does specialization matter? *Agricultural Economics, 32,* 281-296.

Lawson, G. L., Agger, J. F., Lund, M. and Coelli, T. (2004). Lameness, metabolic and digestive disorders, and technical efficiency in Danish dairy herds: a stochastic frontier production function approach. *Livestock Production Science, 91,* 157-172.

Lee, D., Newman, P. and Price, R. (1999). *Decision making in organisations.* Glasgow: Prentice Hall.

Lunneryd, D. (2003). Unique Decision Making with Focus on Information Use - The case of converting to organic milk production. *Acta Universitatis Agriculturae Sueciae. Agraria 405.* Swedish University of Agricultural Sciences.

MacDonald, J,M., O'Donoghue, E.J, McBride, W.D., Nehring, R.F., Sandretto, C.L. and Mosheim, R. (2007). Profits, Costs, and the Changing Structure of Dairy Farming. *Economic Research report no 47.* USDA Economic Research Service, Washington D.C., U.S.

Meeusen, W. and van den Broeck, J. (1977). Efficiency Estimation from Cobb-Douglas production functions with composed errors. *International Economic Review, 18,* 435-444.

Nuthall, P. L. (2001). Managerial ability - a review of its basis and potential improvements using psychological concepts. *Agricultural Economics, 24,* 247-262.

Öhlmér, B., Olson, K. and Brehmer, B. (1998). Understanding farmers' decision making proceses and improving managerial assistance. *Agricultural Economics, 18,* 273-390.

Öhlmér, B. and Lönnstedt, L. (2004). Design of Economic Information. A Pilot Study of Accounting information in Decision-making Processes. *Food Economics-Acta Agric. Scand., Sect. C, 1,* 222-231.

Ondersteijn, C. J. M, Giesen, G. W. J. and Huirne, R. B. M. (2003). Identification of farmer characteristics and farm strategies explaining changes in environmental management and environmental and economics performance of dairy farms. *Agricultural Systems, 78,* 31-55.

Oude Lansink A., Pietola, K. and Bäckman, S. (2002). Efficiency and productivity of conventional and organic farms in Finland 1994-1997. *European Review of Agricultural Economics, 29,* 51-65.

Pfeffer, J. and Salancik, G. R. (1978). *The External Control of Organizations. A resource dependence perspective.* New YorkHarper and Row, Publishers.

Porter, M. (1980). *Competitive Strategy.* New York: The Free Press.

Reinhard, S., Lowell, C. A. K. and Thijssen, G. J. (2000). Environmental efficiency with multiple environmentally detrimental variables; estimated with SFA and DEA. *European Journal of Operational Research, 121,* 287-303.

Resti, Andrea (1997). Evaluating the cost-efficiency of the Italian Banking System: What can be learned from the joint application of parametric and non-parametric techniques. *Journal of Banking and Finance, 21,* 221-250.

Rockart, J. F (1979). Chief executives define their own data needs. *Harward Business Review. March-April,* 81-92.

Rougoor, C. W., Trip, G., Huirne, R. B. M. and Renkema, J. A. (1998). How to define and study farmers' management capacity: theory and use in agricultural economics. *Agricultural Economics, 18,* 261-272.

Sharma, K. R, Leung, P. and Zaleski, H. M. (1999). Technical, allocative and economic efficiencies in swine production in Hawaii: a comparison of parametric and nonparametric approaches. *Agricultural Economics, 20,* 23-35.

Shingfield, K.J., Jaakkola, S. and Huhtanen, P., 2002. Effect of forage conservation method, concentrate level and propylene glycol on diet digestibility, rumen fermentation, blood metabolite concentrations and nutrient utilization of dairy cows. *Animal Feed Science and Technology, 97,* 1-21.

Simar, L. and Wilson, P. W. (2007). Estimation and inverence in two-stage, semi-parametric models of production processes. *Journal of Econometrics, 136,* 31-64.

Simon, H. A. (1997). Administrative Behavior. New York: The Free Press.

Sorensen, J. T. and Ostergaard, S., 2003. Economic consequences of postponed first insemination of cows in a dairy cattle herd. *Livestock Production Science, 79,* 145-153.

Statistics Sweden. (2002). *Jordbruksstatistisk årsbok 2002.* Statistics Sweden, Örebro, Sweden.

Statistics Sweden (2006). *Jordbruksstatistisk årsbok 2005.* Statistics Sweden, Örebro, Sweden.

Steinbock, L. Näsholm, A., Berglund, B., Johansson, K. and Philipsson, J. (2003). Genetic Effects on Stillbirth and Calving Difficulty in Swedish Holsteins at First and Second Calving. *Journal of Dairy Science, 86,* 2228-2235.

Tauer, W. L. (1993). Short-run and long-run efficiencies of New York dairy farms. *Agricultural and Resource Economics Review, 22,* 1-9.

Tauer, L. W. and Mishra, A. K. 2006 Can the small dairy farm remain competitive in US agriculture? *Food Policy, 31,* 458-468.

Trip, G., Thijssen, G. J., Renkema, J. A. and Huirne, R. B. M. (2002). Measuring managerial efficiency: the case of commercial greenhouse owners. *Agricultural Economics, 27,* 175-181.

Wilson, P., Hadley, D and Asby, C. (2001). The influence of management characteristics on the technical efficiency of wheat farmers in eastern England. *Agricultural Economics, 24,* 329-338.

Wilson, P., Hadley, D., Ramsden, S and Kaltsas, I. (1998). Measuring and Explaining Technical Efficiency in UK Potato Production. *Journal of Agricultural Economics, 49,* 294-305.

In: Agricultural Systems: Economics, Technology and Diversity ISBN 978-1-60692-025-1
Editor: Oliver W. Castalonge

Chapter 2

DOES GIBRAT'S LAW HOLD AMONGST DAIRY FARMERS IN NORTHERN IRELAND?

***Philip Kostov*[1], *Myles Patton*[2], *Joan Moss*[2] and *Seamus McErlean*[3]**
[1]Lancashire Business School, University of Central Lancashire, Preston, UK
[2]Agri-Food and Biosciences Institute, Belfast, UK
[3]Department of Agricultural and Rural Development in Northern Ireland, Belfast, UK

ABSTRACT

This chapter tests whether the Law of Proportionate Effects (Gibrat, 1931), which states that farms grow at a rate that is independent of their size, holds for the dairy farms in Northern Ireland. Previous studies have tended to concentrate on testing whether the law holds for all farms. The methodology used in this study permits investigation of whether the law holds for some farms or all farms according to their size. The approach used avoids the subjective splitting of samples, which tends to bias results. Additionally we control for the possible sample selection bias. The findings show that the Gibrat law does hold except in the case of small farms. This is in accordance with previous findings that Gibrat's law tends to hold when only larger farms are considered, but tends to fail when smaller farms are included in the analysis. Implications and further extensions, as well as some alternatives to the proposed methodology are discussed.

Key words: Gibrat's law, quantile regression, integrated conditional moments test

JEL codes: C12, C14, O49.

INTRODUCTION

Farm structural change and the inter-related issue of farm growth continue to attract the interest of agricultural economists, academics and policy makers, because of the wide-ranging implications for agricultural output, efficiency and inevitably the economic welfare of rural communities. Given the sweeping changes that have occurred in the recent round of CAP Reform it is important that the factors affecting structural change in the farm sector are understood. This study will examine some of these factors using data from the period 1997 to 2003. During this period, farm incomes reached very low levels (see Figure 1) and this has certainly put pressure on the structure of the farm sector.

Two inter-related components of farm structural change are entry/exit to the farm sector and growth/decline in continuing farms (Weiss, 1999). Historically, there has been a tendency for academics to consider these two components separately. Studies that have investigated the factors influencing farm exit include Kimhi and Bollman (1999) and Glauben *et al.* (2003). A large number of studies have examined the factors affecting farm growth (e.g. Upton and Haworth, 1987; Shapiro *et al.*, 1987; Clarke *et al.*, 1992; and, Bremmer *et al.*, 2002). However, it has been argued that examining the growth of continuing farms only, whilst ignoring exiting farms in the analysis, runs the risk of biasing results (Weiss, 1999). Consequently, several more recent studies have considered both farm survival (the opposite of exiting) and farm growth (see Weiss, 1998 and 1999; and, Key and Roberts, 2003).

In a seminal paper that describes the Law of Proportionate Effects, Gibrat (1931) provides the starting point for most previous studies of farm growth. He proposed that the growth rate of firms is independent of their initial size at the beginning of the period examined. Gibrat's law proposes that growth is a stochastic process resulting from the random operation of many independent factors. This stochastic process easily generates theoretical farm size distributions (log-normal) that are skewed and similar in shape to the farm size distributions that are observed empirically. Furthermore, this stochastic process also means the variance of the distribution increases over-time mirroring observed increases in empirical concentration measures.

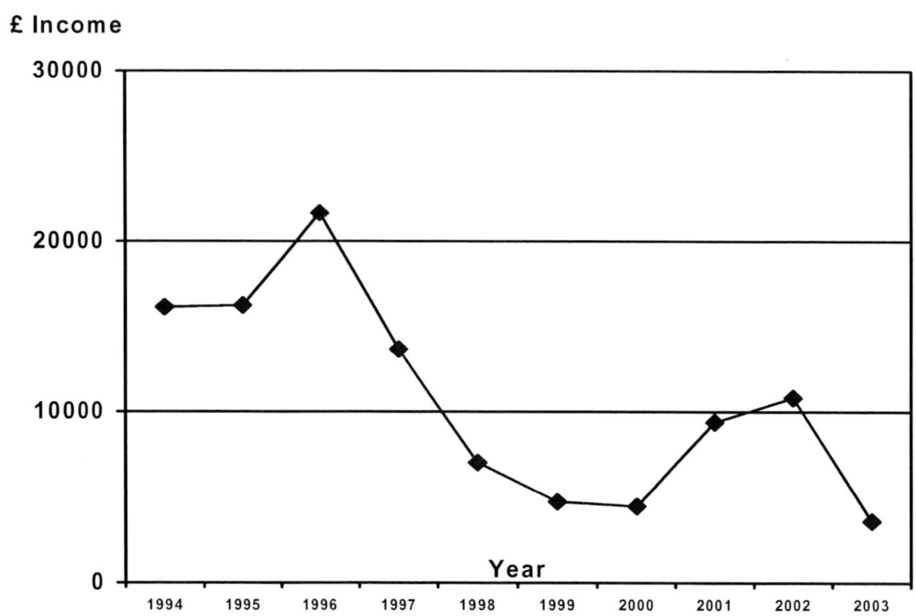

Figure 1. Net farm income of Northern Ireland dairy farms.

It is unsurprising, therefore, that Gibrat's Law, which is not inconsistent with the assumption of constant returns to scale, continues to provide the basic foundation for most studies of farm growth.

Gibrat's law, however, has been subject to some criticism and theoretical justifications for its rejection can be found in the literature, e.g., returns to scale among smaller farms (Weiss, 1999). A number of studies have identified a range of systematic factors that influence farm growth and these should be explicitly considered in any model of farm growth, rather than being subsumed within the random stochastic process implied by Gibrat's law. In light of these recent considerations this paper test Gibrat's law using some determinants of farm growth and survival using farm census data for dairy farms in Northern Ireland. The analysis permits examination of Gibrat's law and considers the influence of such factors as profitability, farm type and farmer associated characteristics on farm growth and survival. An outline of the paper is as follows. A review of previous studies is presented in the next section. The data set and the methodology used in the analysis are outlined in Section 3. Model results are presented in Section 4. These results are discussed in Section 5 and some conclusions are drawn in Section 6.

Literature Review

There is a wide and extensive literature investigating the growth of firms (for a review see Sutton, 1997; and, Lotti et al., 2003). In comparison the number of studies focusing on farm growth is more limited. The approach used in most studies of farm growth has been to test Gibrat's law. Many of these studies appear to have reached different conclusions. Studies by Weiss (1998, 1999) and Shapiro *et al.* (1987), based on farm census data, rejected Gibrat's law of proportionate effects for farm growth. These studies found that small farms tend to grow faster than larger farms. However, Upton and Haworth (1987) and Bremmer *et al.* (2002) using FBS data collected in Great Britain and FADN data collected in the Netherlands, respectively, found no evidence to reject Gibrat's law. The data sets used in these two studies exclude very small farms (i.e. farms <8 European Size Units). This may have affected the results obtained because small rapidly growing farms may have been excluded from the analysis. Clark *et al.* (1992) also find no evidence to reject Gibrat's law based on analysis using aggregated data.

An important aspect of the study of farm (size) growth is the definition of farm size. Previous studies have used a variety of different measures of farm size (changes in farm size indicate growth/decline). Measures of farm size proposed in the literature include: acreage farmed, livestock numbers (cow equivalents), total capital value, net worth, gross sales, total gross margins and net income (Allanson, 1992; Clark *et al.* 1992; and, Shapiro *et al.* 1987). Output value measures such as gross farm sales and input value measures such as net worth may be unsatisfactory due to the impact of inflation and changes in relative prices (Weiss, 1998). Physical input measures, such as acres under cultivation and number of livestock, are also problematic since farms are characterised by a non-linear production technology and changes in farm size typically involve changes in factor proportions and changes in technology. Weiss (1998), however, argues that the disadavantages of physical input measures are less that those associated with the value of inputs or outputs and as a result the former should be preferred.

A review of empirical studies indicates that most find evidence to conclude that a range of variables other than size influence farm growth (e.g. Weiss, 1999; and, Bremmer *et al.*, 2002). These other explanatory variables that have been identified in the literature can be divided into two sub-groups, namely, farmer-associated characteristics and farm specific factors. Weiss (1999) identified the following farmer-associated characteristics: off-farm employment, age, gender, level of education, family size, age profile of the family, succession information and attitude to risk. Farm (or firm) specific variables that have been suggested as factors influencing growth include: size, solvency, profitability, productivity, farm income, structure, financial performance, input costs, output mix, farm type, mechanisation and location (Bremmer *et al.* 2002; Hardwick and Adams, 2002; Weiss, 1999).

PROBLEM CHARACTERISATION

As with the vast majority of studies of farm growth, Gibrat's law of proportionate effect is used as a starting point for the analysis carried out in this paper. The law states that firm (or farm) growth is determined by random factors, independent of size. This may be tested using the following formula:

$$ln(S_{i,t}) = \beta_1 + \beta_2 ln(S_{i,t-1}) + u_t; \qquad (1)$$

where $S_{i,t}$ denotes the size of the individual holding in time t and u_t is a random disturbance term independent of current or past values of the dependent variable. If $\beta_2 = 1$, then growth rate and initial size are independent and this means that Gibrat's Law is not rejected. If $\beta_2 < 1$, small farms tend to grow faster than larger farms – i.e., the effects of randomness are offset by negative correlation between growth and size. If $\beta_2 > 1$, larger farms tend to grow faster than smaller farms. The above is easily illustrated if one subtracts $ln(S_{i,t-1})$ from both sides of equation (1) above. Then the left hand side logarithmic difference is an approximation of the growth. The right hand side will them be either a random walk, when $\beta_2 = 1$, or a dependent process otherwise.

Equation (1) can be generalised augmenting it by farmer associated characteristics and farm specific variables, e.g., profitability, denoted by the k explanatory variables X_k below as follows:

$$\ln(S_{i,t}) = \beta_1 + \beta_2 \ln(S_{i,t-1}) + \sum_{j=3}^{k+3} \beta_j X_{j-2} + u_t \qquad (2)$$

There are two problems with using linear regression representations such as equation (2). The first is the assumed linear effect of the additional explanatory variables X_k. Weiss (1999) for example applied non-linear functional form for these and detected significant non-linearities. Specifying an ad-hoc non-linear functional form however is not a viable strategy, since it may impact on the final results in an unpredictable way and often there is little or no information on the way these additional variables may impact on farm growth.

This consideration aside, even in the simple model (1), there is an underlying assumption that the Gibrat's law holds (or is violated) globally. It is difficult to ascertain whether for example small farms obey this law as opposed to large farms. It is in principle possible to split the sample into smaller subsamples and locally estimate the relationships. This would however involve some subjective criteria about how to do the latter partitioning further casting doubt on the final results. If we want to test whether Gibrat's law holds for some farms and not for others, the linear regression framework is too restrictive. Such a test can nevertheless be designed using quantile regression methods, implemented in this paper. Alternatives and extensions to the adopted approach are also discussed.

In order to measure farm growth, farm size must be compared between two specific points in time. However, measures of farm growth are meaningful only for surviving farms. Farms exiting between the points in time over which growth is measured are normally excluded from the sample (as non-surviving farms). However, there is a greater probability that slower growing small farms will be non-survivors compared with slower growing larger farms (Weiss. 1999). Thus, if non-surviving (exiting) farms are excluded from the sample, the estimates of β may be biased downward, which may result in incorrect rejection of Gibrat's law, giving the impression that smaller farms tend to grow faster than larger farms (Hardwick and Adams, 2002; Lotti *et al.* 2003; Shapiro *et al.* 1987; Sutton, 1997; and, Weiss, 1999). Ignoring exiting farms in the analysis is known as the problem of sample selection bias. Various options are available to account for selection bias and these are briefly discussed in the methodology section.

DATA

The data set used in this study is based on the 1997 and 2003 farm census for Northern Ireland and a structural survey of farms in Northern Ireland that was conducted in 1997. The farm census provided information for individual farms on farm type, acreage farmed and stock numbers (total standard gross margin for each farm can also be inferred from this data). The 1997 structural survey provided additional information on a range of farmer associated characteristics such as gender, age, management status and time spent working on the farm for a subset of the farms included in the farm census (31 % of dairy farms). The individual farm information from the structural survey was matched to the information from the 1997 farm census.

Matching these data sets yielded a total of 1648 dairy farms in 1997. Of these farms, 112 had exited farming by 2003. Of the remaining 1536 farms, 1290 remained in dairy, while the other 246 moved to cattle and sheep. In this study we are specifically interested in farms which remain in dairying and thus the latter farms are treated as farms which exited the dairy sector. Thus, in total 358 exited the dairy sector between 1997 and 2003 (farms which exited farming altogether plus farms which switched from dairy to cattle and sheep).

The measure of farm size used in this study is the livestock numbers measured in cow equivalents per farm. This measure (unlike e.g., land area) is directly proportionate to the final output of dairy farm. Using the dairy farm sector allows us to avoid complications associated with farm entry and thus simplify the sample selection problem.

The following explanatory variables are employed: (logarithm of the) initial (i.e. in 1997) size - LNCE97; an indicator variable denoting whether the farm holder is also a manager of the farm – MSHOLD, indicator of other gainful activities – HAGA3 and a variable showing the age of the farmer – HAGE1. Note that we use a limited set of conditioning variables, since our purpose is to test Gibrat's Law, rather than provide a comprehensive model of farm growth, which would have involved additional behavioural assumptions and theoretical model.

METHODOLOGY

In the least-squares regression framework the conditional mean function, i.e., the function that describes how the mean of y changes with the covariates x, is almost all we need to know about the relationship between y and x. The crucial aspect about this view is that the error is assumed to have exactly the same distribution irrespectively of the values taken by the components of the vector x. This can be viewed as a pure 'location shift' model since it assumes that x affects only the location of the conditional distribution of y, not its scale, or any other aspect of its distributional shape. If this is the case, we can be fully satisfied with an estimated model of the conditional mean function.

The above described location shift model is however rather restrictive. Covariates may influence the conditional distribution of the response in many other ways: expanding its dispersion (as in traditional models of heteroscedasticity), stretching one tail of the distribution, compressing the other tail (as in volatility models), and even inducing multimodality. Explicit investigation of these effects can provide a more nuanced view of the stochastic relationship between variables, and therefore a more informative empirical analysis. The quantile regression is a method that allows us to do so.

Given a random variable Y and its distribution function F, we denote by

$$Q(\tau) = \inf(y \mid F(y) \geq \tau) \tag{3}$$

the τ th quantile of Y. The sample analogue q of $Q(\tau)$ is called the τ th sample quantile. It may be formulated as the solution of the following optimisation problem, given a random sample (y_n), $n=1;\ldots;N$:

$$\min\left\{ \sum_{\{n|y_n \geq q\}} \tau|y_n - q| + \sum_{\{n|y_n < q\}} (1-\tau)|y_n - q| \right\} \tag{4}$$

There exist a number of alternative quantile regression estimators. Here we will only describe the linear programming type of estimator, since there are asymptotic theory results for it (Koenker and Bassett, 1978). Just as we can define the sample mean as the solution to the problem of minimizing a sum of squared residuals, we can define the median (which is the 50% quantile, i.e., τ =0.5) as the solution to the problem of minimizing a sum of absolute residuals (which follows directly from (4) above).

For any 0 < τ < 1, we denote $\rho_\tau(u) = u(\tau + I_{[u<0]})$, where I[.] is the indicator function. Following Koenker and Bassett (1978) $\rho_\tau(u)$ is usually referred to as a *check* function. The problem may then be formulated as follows:

$$\min \sum_{n=1}^{N} \rho_\tau(y_n - q) \tag{5}$$

which yields a natural generalization to the regression context.

$$\min \sum_{n=1}^{N} \rho_\tau(y_n - \xi(X, \beta)) \tag{6}$$

where $\xi(X, \beta)$ is some parametric function of the covariates. When this is a linear function, the above minimisation procedure is actually a linear programming problem. Then it may be estimated using some form of simplex algorithm. Koenker and d'Orey's (1987, 1993) adaptation of the Barrodale and Roberts (1974) median regression algorithm to general quantile regression is particularly influential. The Barrodale and Roberts approach belongs to the class of exterior point algorithms for solving linear programming problems. Alternatively, Portnoy and Koenker (1997) have shown that a combination of interior point methods and effective problem preprocessing is very well suited for large-scale quantile regression problems. This is the approach used in this paper (it is often referred to as Frisch-Newton method), although the former (Barrodale-Roberts) method yields similar results, which are available from the authors upon request.

It would be beneficial at this point to clarify a fundamental difference between the quantile regression and the mean regression methods. Could we achieve the same result by simply segmenting the response variable into subsets according to its unconditional distribution and then doing least squares fitting on these subsets? Clearly, this form of truncation on the dependent variable would yield disastrous results in the present example. In general, such strategies are doomed to failure for all the reasons so carefully laid out in Heckman (1979). It is thus worth emphasizing that even for the extreme quantiles all the sample observations are actively used in the process of quantile regression fitting.

It is of course possible to construct local quantile regression estimates using some sort of segmentation (see Knight et al., 2002). Some preliminary results about the conditions where local quantile regression is useful are outlined in Costinot et al. (2000). However we will not pursue this option for reasons given further below.

There are several useful properties of the quantile regression approach. Above we have described the quantile regression for a given quantile. If however one takes the whole range of quantiles, a picture of the overall distribution emerges. Note that in the latter case we obtain a variable coefficients model. In contrast to most variable coefficient methods which usually assume coefficients independence however, in the quantile regression setting, the coefficients are functionally dependent. In the light of the farm growth problem, this is evidently a desirable property, The determinant of the farm growth for slightly different sizes of farms are related in a quantile regression context, while paradoxically they will be assumed independent in most other variable coefficients models. Even the simplest linear quantile regression we adopt here produces a rather flexible non-linear model. Note furthermore that the non-linearities are explicitly formulated with regard to the dependent variables, i.e., with regard to the farm size, which is exactly what ids necessary for testing Gibrat's law. Nevertheless, the quantile regression has to be viewed as a workable approximation to a possibly more general non-linear model. The availability of pointwise convergence results for the quantile regression estimates facilitates the analysis and inference compared to other non and semi-parametric methods.

The potential problem of bias due to sample attrition is known in the literature as a sample selection problem. Its initial description is due to Heckman (1979) who devised a two step procedure for controlling it. The Heckman procedure consists of estimating at step one a survival model. This is typically a probit (although a logit can be used alternatively) equation on the probability of farm survival from the complete sample (including surviving and non-surviving farms). This equation is subsequently used to obtain an additional variable, where the values represent the inverse Mill's Ratio for each observation. In step-two, the additional variable is introduced as a correcting factor into the least squares regression that is based upon a sample that excludes non-surviving farms. The probit model, used in the first step typically has the same explanatory variables as the main equation, though this is not mandatory and variables that are only relevant to the farm survival may be included, as well as some of the variables included in the main equation may be dropped. The Heckman procedure assumes joint normality of the error terms in the two equations. The latter distributional assumption, which can also be employed to construct a more efficient Full Information Maximum Likelihood (FIML) estimator that jointly estimates both equations, however can have serious implications on the robustness of the final results when it is violated. Therefore various alternative estimators have been suggested to circumvent the problem of inadequate distributional assumptions. These can be broadly described as semi-parametric model selection methods.

The problem of sample selection in mean regression model can be broadly defined as problem of the distributional assumptions, which can be controlled for. This is basically done by various methods to relax the parametric specifications employed in the seminal work of Heckman (1979).

The sample selection for quantile regression however remains a challenging and still under-researched problem. Buchinsky (1998, 2001) provided some important contributions to this issue. Unfortunately the method Buchinsky (1998,2001) used in the selection step, namely the Ishimura's (1993) semiparametric least squares requires that the selection equation includes at least one covariate that is not included in the main equation. This condition is difficult to ensure with the available data set. Therefore a different strategy is followed in this paper. On the first step we estimate an ordinary probit selection equation similarly to Heckman (1979) and from there derive the bias correcting factor (i.e. the inverse Mill's ratio). In the second step a linear quantile regression is performed instead of the mean regression by additionally including the derived correcting factor. The resulting model is tested for model correctness, which also validates the sample selection step.

The last piece of the jigsaw is therefore to identify appropriate model validity test, applicable to the quantile regression. What is needed is a test on validity of the functional form. Up to our knowledge there are only two appropriate candidates for this. The first is the Zheng' (1998) approach based on weighted kernel regression estimation and the other one is an extension to the Bierens and Ploberger's (1997) Integrated Conditional Moment (ICM) test for the quantile regression case due to Bierens and Ginther (2001). They discuss explicitly only the median case, but the necessary modifications for the general quantile regression case are provided in an appendix. We have chosen the latter due to some desirable properties, such as boundness of the test statistic, good local power and relative conservatism of the test statistic. We briefly describe the ICM test below:

Let us have the following expectation model:

$$E\left(y_j \mid x_j\right) = g\left(x_j, \beta\right) \tag{7}$$

then the ICM statistic

$$\hat{T}_{ICM} = \frac{\int \left|\hat{z}(\xi)\right|^2 d\mu(\xi)}{\int \hat{\Gamma}(\xi,\xi) d\mu(\xi)} \tag{8}$$

with $\hat{z}(\xi) = \frac{1}{\sqrt{n}} \sum_{j=1}^{n} \hat{u}_j \omega\left(\xi' \Phi\left(x_j\right)\right)$, $\hat{u}_j = y_j - g\left(x_j, \hat{\beta}\right)$ being weighted non-linear residuals, $\Phi(.)$ - bounded one to one mapping and $\omega(.)$ - appropriate weight function, can be used to test the null hypothesis that the probability of the expectation given above in (7) is one, against the alternative of being the less than one. In simple words, this amounts to testing that the expectational model (7) is the right one. Practical implementation of the ICM test involves choice for the weight function and the bounded mapping as well as some computational issue surrounding the computation of the two integrals in (8). For brevity of the exposition we will not discuss these here. Full details on the procedure, which roughly speaking follows Bierens and Ginther (2001)[1], are available from the authors.

There is another point to note. Since we are computing a two-step estimator, the confidence intervals obtained during the quantile regression estimated at the second step will not be valid even asymptotically. The reasons for this are similar to the instrumental variables case and the endogeneity problems. In the sample–selection case analysed here, in general the quantile regression estimated at the second step is actually:

$$Q(\tau) = g\left(X, \beta_\tau\right) + k_\tau \left[f(.)/\left(1 - F(.)\right)\right] + \varepsilon \tag{9}$$

The additional regressor $f(.)/\left(1-F(.)\right)$ can be interpreted as a 'non-selection hazard', that corrects for the effect of the sample selection bias. The *pdf* $f(.)$ and the *cdf* $F(.)$ are obtained from the auxiliary model estimated at step one. Note that (9) is the expression when the auxiliary model models non-selection. If the auxiliary model is the one of selection the

[1] They do not provide sufficient information on the exact quantile regression estimation algorithm they use or the method of obtaining standard errors, thus full replication was not possible.

natural candidate for the non-selection hazard would be $f(.)/F(.)$. In the case of symmetric distributional assumptions used in the auxiliary model (such as the Gaussian used in the probit model), modelling selection or non-selection would be equivalent. This would not however be the case for semiparametric estimators, and thus it is preferable to use the non-lecetion model in the first step for such cases. The "inverse Mill's ratio" is the representation of the above non-selection hazard obtained from the normal distribution (or the logistic, if a logit model was estimated instead of probit). If one estimates a non-parametric model at step one the empirical *pdf* and *cdf*, obtained from it could be used.

The Heckman type of procedure ignores the correlation between the residuals in the first and second step models. Since the first-step residuals are used to compute the non-selection hazard, the latter would therefore be correlated with the residuals in the second-step equation. With a probit equation estimated in step one, the standard errors in the following linear regression can be adjusted (see Heckman, 1979) obtain asymptotically valid standard errors. An alternative (applicable with any pair of parametric models for the two steps) is to estimate jointly both equations by Full Information Maximum Likelihood, which by accounting for the correlation between the error terms should yield consistent standard errors. Note however that this is not applicable in this case because the quantile regression is a semi-parametric method, which does not make any distributional assumptions. To obtain valid standard errors , which are necessary for testing the Gibrat's law, we adopt a different approach. We consider both equations as a whole and bootstrap the whole two-stage estimator (instead of only the QR estimates). An adapted quantile regression implementation of the XY (also known as pair, or case) bootstrap is used for this purpose.

RESULTS

The first-step estimates are presented in Table 1. Note that since we use probit model, and thus the choice of auxiliary criterion is not instrumental, we model selection (i.e. we use 1 for farms that remain in dairying)[2]. This leads to a more natural interpretation of the estimated coefficients, if this is necessary. We use the same variables as in the subsequent quantile regression model. The farm size is important determinant of the farm exits, in that larger farms are more likely to remain in dairying. Owing to this if the selection process was not controlled for, one would have obtained biased estimates in the main equation.

The discussion in the methodology section concentrated on estimation and testing of a quantile regression model at a given quantile. It is clear that the adopted framework for testing the Gibrat's law involves estimating multiple quantile regression models. It is in principle possible to estimate the whole quantile process (i.e. estimating a quantile regression for every observation, in this case 1290 models).

[2] Correspondingly the additional variable used in the subsequent quantile regressions is f(x)/F(x).

Table 1. First-step equation estimates

Variable	Coefficient	Standard Error
Constant	0.197	0.092
LNCE97	0.180	0.013
MSHOLD	-0.043	0.023
HAGA3	-0.009	0.050
HAGE1	-0.003	0.001

To simplify the process however only a subset of quantile regression models is estimated. This subset consist of all percentiles excluding the lower and the upper 9%. In other words the 81 regression models for the 0.10, 0.07, 0.89, and the 0.90 quantiles were estimated. The reason we exclude the extreme quantiles is that the conventional quantile regression estimates for these are unreliable. Asymptotic theory and estimation methods for extreme quantiles are developed in Chernozhukov (2000a,b) and Chernozhukov and Umantsev (2001). The main interest of the current paper lies in the overall distribution of the estimated on the logged size in the initial period (1997) and thus we will ignore the extreme quantiles.

Plotting the corresponding estimates for the same parameter across the quantile range provides a useful graphical device to informally ascertain the scale invariance hypothesis. Strict formal tests on this are available and results from such tests can be provided by the authors upon request. The conventional approach to such tests (Koenker and Xiao, 2000) uses the Khmaladze (1981) transformation and introduces additional computational burden. Although we omit it here for brevity and simplicity, rigorous modelling practices would require one to implement such tests. Graphical representation of the overall quantile regression is presented on Figure 2.

The main point of interest here however, is the way the estimate of the logged lagged size varies over the quantile range. For this reason we will not comment on the overall results and will focus our attention to this particular coefficient. The estimates for the coefficient of the initial farm size from the estimated set of quantile regressions are presented on Figure 3, together with the associated 95% confidence intervals. Where the horizontal line drawn at the value of 1 falls within the range defined by these confidence intervals we may say with 95% confidence level, that the Gibrat law holds. These results suggest the following. The Gibrat's law speaking holds except for the small (up to the 0.16 quantile) dairy farms. These smaller farms grow slower than the rest of the sector. Interestingly the coefficient estimate declines for the largest farms, but this decline is not statistically significant. If we also include the extreme quantiles, the largest farms do show slower growth than the rest (results available from the authors). Nevertheless such a result is difficult to verify, since the ICM test statistic is unreliable at the extreme quantiles.

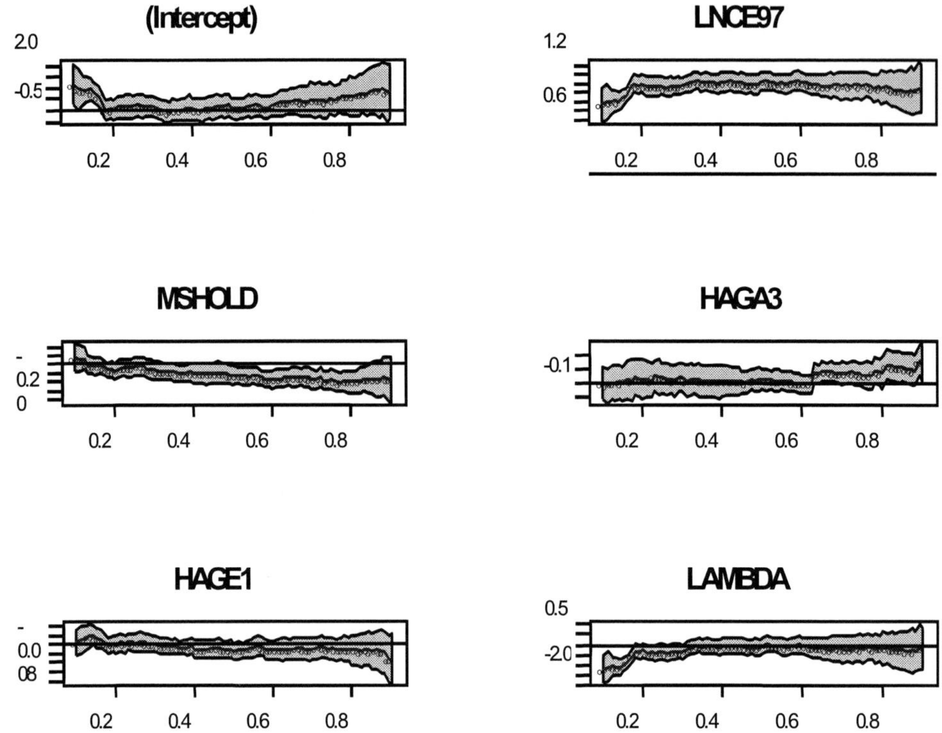

Figure 2. Graphical representation of the overall quantile regression results.

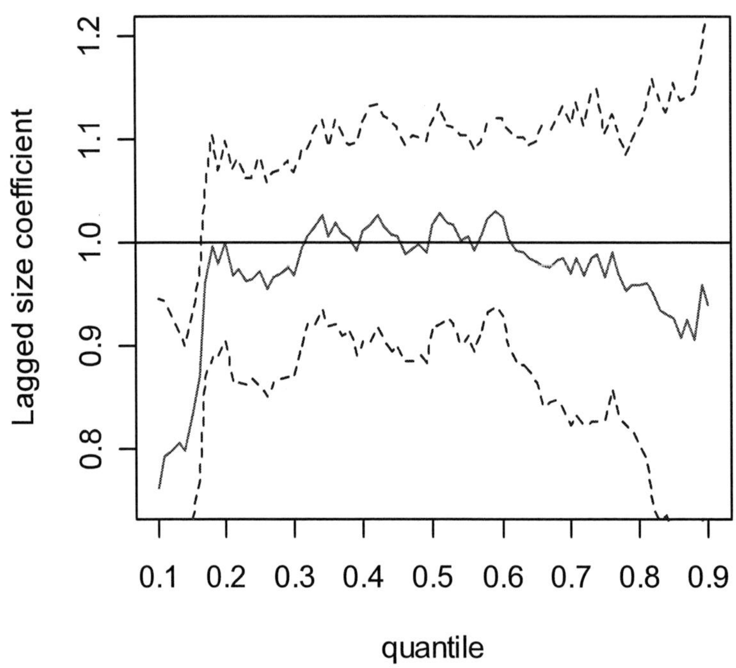

Figure 3. Quantile regression estimates on the lagged size coefficient.

The smaller growth in the segment of smaller dairy farms is in concordance with farm growth results from Census and farm business surveys data, since, the latter are generally based on larger farms, and thus will tend to support the Gibrat's law as opposed to the latter where the peculiarity of the smaller farms would lead to its rejection. The use of dairy sector data is advantageous in that it reduces the possibility for heterogeneity problems due to different production technologies and farm types being pooled together. McErlean et al. (2004) argue that this heterogeneity needs to be dealt with by explicitly modelling the different farm types instead of using dummies for them. Note that the possible effects of such heterogeneity will manifest themselves in terms of heteroscedasticity problems. The distribution of the ICM test we use to test the model validity however is not affected by neglected heteroscedasticity (Bierens and Ploberger, 1997).

Due to the considerable computational burden of estimating the ICM test statistic (details on the exact procedure available upon request) we only estimate it for the quantiles from 0.1 through to 0.9 with 0.1 steps). The ICM test results are presented in Table 2.

The reason for using several values for c is as follows. The ICM statistic is a ratio of two probability measures estimated over a hypercube whose dimensions are $2c$. (i.e., in the intervals $[-c, c]$). In principle asymptotically any choice for c is equivalent. In principle however, this choice may have dramatic effects on the small sample properties of the test.

Table 2. ICM test results

Quantile	c=1	c=3	c=5	c=10	c=15	c=20
0.1	0.050	0.247	0.363	0.674	0.698	0.665
0.2	0.110	0.531	0.628	0.746	0.791	0.817
0.3	0.119	0.500	0.521	0.718	0.827	0.810
0.4	0.364	1.683	1.580	1.485	1.336	1.150
0.5	0.338	1.466	1.409	1.655	1.348	1.148
0.6	0.398	1.571	1.231	1.229	1.115	1.025
0.7	0.389	1.076	0.882	0.712	0.755	0.834
0.8	0.179	0.758	0.639	0.797	0.815	0.959
0.9	0.068	0.264	0.399	0.674	0.734	0.733

Critical values (Bierens and Ploberger, 1997):
10%3.23.
5% 4.26.

In general too small or too large values will reduce the power of the test. (see Bierens and Ginther, 2001 for a more detailed discussion on this in the quantile regression case). Therefore a range of such values was used to estimate the ICM test. All test statistics estimated fail to reject the null of validity of the estimated quantile regression. The results for c=0.1 similarly to Bierens and Ginther (2001) are probably spuriously low. Nevertheless the range of values for the hypercube dimension is rather extensive (as a comparison Bierens and Ginther (2001) only use values of 1, 5 and 10) and everywhere the ICM test statistic is well below the critical values. This provides conclusive evidence in support of the estimated quantile regression model and its conclusions.

CONCLUSION

Previous studies show that Gibrat's law tends to hold when only larger farms are considered, but tends to fail when smaller farms are included in the analysis. This study is based on a data set that covers the full range of dairy farm sizes in Northern Ireland. The analysis takes account of possible bias due to exiting farms. Our results indicate that the farm growth does not depend on initial size, except for the smaller farms. Small Northern Ireland dairy farms relying on family labour, probably experience resources shortage and have insufficient funds to expand under milk quota restrictions. On the other hand the largest farms seem to grow slower than the rest, hinting a possible a saturation effect, but the latter is inconclusive. The relatively small average dairy farm size in Northern Ireland may explain such an effect.

The use of rather homogeneous data set consisting of only dairy farms have prevented some complications such as possible heterogeneity, but the general approach outlined in the paper is readily applicable to more complex data sets. In such cases the simple probit sample selection step may not be appropriate and alternative semi-parametric formulations may be used instead. The linear quantile regression proved to be sufficient to describe the growth process in the NI dairy sector. In some other cases however the linear assumption may be inadequate. Then nonlinear and non-parametric versions of the quantile regression could be employed instead.

REFERENCES

Allanson, P. (1992). Farm size structure in England and Wales 1939-89. *Journal of Agricultural Economics* 43(2), 137-148.

Barrodale, I., and Roberts, F. (1974). Solution of an overdetermined system of equations in the L1 norm. *Communications of the ACM* 17, 319-320.

Bierens H. J. and Ploberger, W. (1997). Asymptotic theory of integrated conditional moment tests. *Econometrica* 65, 1129-1151.

Bierens, H. J. and Ginther, D. K. (2001). Integrated conditional moment testing of quantile regression models. *Empirical Economics* 26, 307-324.

Bremmer, J., Oude Lansink, A.G.J.M., Olson, K.D., Baltussen, W.H.M., and Huire, R.B.M. (2002). Analysis of farm development in Dutch agriculture and horticulture. Paper presented at 13th Congress of the International Farm Management Association (IFMA) in National Sports and Conference Centre Papendal, Wageningen, The Netherlands, July 7-12, 2002.

Buchinsky, M. (1998). The dynamics of changes in the female wage distribution in the USA: a quantile regression approach. *Journal of Applied Econometrics* 13, 1—30.

Buchinsky, M. (2001). Quantile regression with sample selection: estimating women's return to education in the U.S. *Empirical Economics* 26(1), 87-113.

Chernozhukov, V. (2000a). Conditional extremes and near-extremes, MIT Dept. of Economics Working Paper No. 01-21.

Chernozhukov, V. (2000b). Conditional extremes and near-extremes: concepts, asymptotic theory and economic applications, unpublished PhD Dissertation, Stanford University.

Chernozhukov, V. and Umantsev, L. (2001). Conditional value-at-risk: aspects of modelling and estimation. *Empirical Economics* 26(1), 271 – 292.

Clarke, J.S., Fulton, M. and Brown, D.J. (1992). Gibrat's law and farm growth in Canada. *Canadian Journal of Agricultural Economics* 40, 55-70.

Davies, S. and Lyons, B. (1982). Seller concentration: the technological explanation and demand uncertainty. *Economic Journal* 92, 903-919.

Gibrat, R. (1931). *Les Inegalites Economiques*. Librairie du Recueil Sirey. Paris.

Glauben, T., Tietje, H. and Weiss, C. (2003). Agriculture on the move: exploring regional differences in farm exit rates. Working Paper EWP 0308. Dept. of Food Economics and Consumption Studies, University of Kiel.

Hardwick, P. and Adams, M. (2002). Firm size and growth in the UK life insurance industry. *The Journal on Risk and Insurance* 69(4), 577-593.

Heckman, J. J. (1979). Sample selection bias as a specification error. *Econometrica* 47(2), 153-161.

Key, N. and Roberts, M.J. (2003). Government payments and structural change in agriculture. Paper presented at the Annual Meeting of the AAEA, Montreal, Quebec, July 27-30, 2003.

Kimhi, A. and Bollman, R. (1999). Family farm dynamics in Canada and Israel: the case of farm exits. *Agricultural Economics* 21, 69-79.

Knight, K, Bassett, G. W. and Tam, M. (2002). Quantile models and estimators for data analysis. *Metrika* 55, 17-26.

Koenker, R. and Basset, G. (1978). Regression quantiles. *Econometrica* 46, 33-50.

Koenker, R. and Xiao, Z. (2000). Inference on the quantile regression process. *Econometrica* 81, 1583–1612.

Khmaladze, E. (1981). Martingale approach in the theory of goodness-of-fit tests. *Theory of Probability and its Applications* 26, 240–257.

Lotti, F., Santarelli, E. and Vivarelli, M. (2003). Does Gibrat's Law hold among young, small firms?. *Journal of Evolutionary Economics* 13(3), 213-235.

McErlean, S., Moss, J. Patton, M. Kostov, P. and Jack, C. (2004). Investigating the determinants of farm growth and survival, Paper presented at the AES annual conference, Imperial College, London, April.

Shapiro, D., Bollman, R.D. and Ehrensaft, P. (1987). Farm size and growth in Canada. *American Journal of Agricultural Economics* 69, 477-483.

Sutton, J. (1997). Gibrat's Legacy. *Journal of Economic Literature* 35, 40-59.

Upton, M. and Haworth, S. (1987). The growth of farms. *European Review of Agricultural Economics* 14, 351-366.

Weiss, C.R. (1998). Size, growth, and survival in the upper Austrian farm sector. *Small Business Economics* 10, 305-312.

Weiss, C.R. (1999). Farm growth and survival: econometric evidence for individual farms in Upper Austria. *American Journal of Agricultural Economics* 81, 103-116.

Zheng J. X. (1998). A consistent nonparametric test of parametric regression models under conditioning quantile restrictions. *Econometric Theory* 14, 123-138.

In: Agricultural Systems: Economics, Technology and Diversity ISBN 978-1-60692-025-1
Editor: Oliver W. Castalonge

Chapter 3

SHANNON ENTROPY OF AMMONIA VOLATILIZATION FROM FERTILIZED AGRICULTURAL SOILS

G. D. Liu[1], Y. C. Li*[1] and A. K. Alva[2]

[1]Department of Soil and Water Sciences, Tropical Research and Education Center, University of Florida, 18905 SW 280th St., Homestead, FL 33031, USA

[2]Vegetable and Forage Crops Research Laboratory, USDA-ARS, 24106 N. Bunn Rd., Prosser, WA 99350, USA

ABSTRACT

The economic loss of ammonia (NH_3) volatilization from chemical N fertilizers applied to farmlands worldwide is 11.6 billion US dollars per year. The economic impact of negative environmental effects resulted from NH_3 volatilization, i.e., formation of potent greenhouse gas (N_2O) and $PM_{2.5}$, is difficult to estimate but enormous. The Shannon's Information Theory was applied to the data collecteddd from our previous study using either ammonium sulfate [$(NH_4)_2SO_4$], ammonium nitrate (NH_4NO_3), potassium nitrate (KNO_3), or urea applied to four agricultural soils, i.e., Biscayne Marl Soil (BMS) and Krome Gravelly Loam (KGL) from Florida and Quincy Fine Sand (QFS) and Warden Silt Loam (WSL) from Washington, at two soil water regimes (20% and 80% field capacity, FC) and incubated at either 11, 20, or 29°C. Shannon rate (R_s g N ha^{-1} bit^{-1}) was defined as N-loss per unit area per Shannon entropy via NH_3 volatilization from the fertilized soils. The results showed that the R_s values across the four soils were 3-fold greater at 20% FC than that at 80% FC soil water regime. The BMS and KGL soils depicted similar R_s values, i.e., 2362.5 and 2378.9, while those for the QFS and WSL soils were only 1079.0 and 851.1 g N ha^{-1} bit^{-1}, respectively. The R_s values were 4178.6, 2863.3, 1502.3, and 21.7 g N ha^{-1} bit^{-1} for urea, $(NH_4)_2SO_4$, NH_4NO_3 and KNO_3, respectively. The environmental friendliness of either the tested soils, fertilizers, or soil water regimes ,etc. were based on the R_s values, discussed.

* Corresponding author (yunli@ufl.edu).

INTRODUCTION

Ammonia (NH_3) volatilization mainly from agriculture such as fertilized soils and dairy farms causes two serious problems: agricultural and environmental (Liu et al., 2007). In the agricultural aspect, nitrogen (N) loss from applied fertilizers via NH_3 volatilization leads yield loss and increases costs in crop production. In fact, NH_3 volatilization from N fertilizers used for agricultural production reduces N uptake. The annual world-wide economic loss is US$ 11.6 billion due to NH_3 volatilization from chemical N fertilizers applied to farmlands (FAO, 2001). This chapter will focus more on the environmental aspect.

In the environmental aspect, volatilized NH_3 may contribute to acid rain, global warming and air pollution health problems associated to. Volatilized NH_3 is the only natural alkaline gas and the third most abundant N gas (after N_2 and N_2O) in the earth's atmosphere (Asman et al., 1982; Schlesinger and Hartley, 1992). Volatilized NH_3 has a short residence time of only about 10 days in the atmosphere since its rapid conversion to nitrous oxide (N_2O) (Dentener and Crutzen, 1994) and/or ammonium (NH_4^+), and the deposition of volatilized NH_3 onto soil and water surfaces (Fowler et al., 1997; Aneja et al., 1998). There is an annual flux of about 7.5×10^7 MT (metric tons) of N derived from the global sources of NH_3 volatilized into the atmosphere (Schlesinger and Hartley, 1992).

Most of the volatilized NH_3 can be deposited *in situ* (within ca 50 km from the source) or *ex situ* (ca 400 km from the source) by either dry deposition or wet deposition (Duce et al., 1991; Schlesinger and Hartley, 1992; Ferm, 1998; Warneck, 1999). Schlesinger and Hartley (1992) estimated that 76% of volatilized NH_3 was annually deposited onto water or soil surfaces. This deposited NH_3 causes soil and water body acidification, eutrophication and forest dieback (Van Breemen et al., 1982).

The volatilized NH_3 also exacerbates global climate change. It was estimated that 4%of the globally volatilized NH_3 (Dentener and Crutzen 1994) can be oxidized by OH radicals and NO_2 mainly in the tropics (Finlayson-Pitts and Pitts, 2000). A fraction of the oxidized NH_3 is transformed to N_2O and this can count for 5% of the global N_2O emission (Ferm, 1998). N_2O is a potent greenhouse gas and approximately 310-fold more powerful than CO_2 in trapping heat in the atmosphere (Finlayson-Pitts and Pitts, 2000).

The remainder of the volatilized NH_3 reacts with acid gases such as sulfur dioxide (SO_2) generated from fossil fuel combustion; and these reactions provide a major portion of the ambient fine particulate matter called $PM_{2.5}$ (the fraction of aerosol particles with an aerodynamic diameter less than 2.5 microns) (Finlayson-Pitts and Pitts. 1986). $PM_{2.5}$ particles are harmful to human health because they can be inhaled and can penetrate into the gas-exchange region of the lung (Brunekreef and Holgate, 2002). Therefore, $PM_{2.5}$ particles cause numerous health problems including asthma, bronchitis, acute and chronic respiratory symptoms such as shortness of breath and painful breathing, and premature death (Marcazzan et al., 2001).

Shannon's information entropy (or simply Shannon entropy) was developed by Claude Shannon (Shannon 1948; Shannon and Weaver, 1949). Shannon entropy is different from thermodynamic entropy. The latter means the thermodynamic homogeneity of a system that limits the ability to extract mechanical work. The former is, in statistics, the amount of information that is lacking in a random variable. This concept has been applied to a broad variety of disciplines (Fleming, 2007) including: physics (Abe, 1997), chemistry (Bonchev and Rouvray, 2005), genomics (Marenduzzo et al., 2006), molecular biology (Frederick et al., 2007), soil science (Martin and Rey, 2000), hydrology (Kawachi et al., 2001), and ecology (Hill, 1973). In ecology, for example, Shannon entropy is used as a measure of biodiversity in the study of ecology. However, there is little information on the application of Shannon's theory to NH_3 volatilization. Therefore, the application of Shannon's theory to NH_3 volatilization from agricultural soils has been investigated in this study.

SHANNON ENTROPY OF AMMONIA VOLATILIZATION

According to Shannon's theory (1948), information entropy was developed as a successful device to gauge information content in communication.

$$H = -K\sum_{i=1}^{n} p_i \log p_i \tag{1}$$

where H denotes the random variable, p_i is the probability of the ith outcome of a discrete random variable, $i = 1, 2, \ldots$, n, n is the number of possible outcomes, K is a positive constant depending on a choice of a unit of measure (Chakrabarti and Chakrabarty, 2005). For example, K equals 1 if digit 2 is taken as the base of the logarithm. We have not specified the base for the logarithm to be taken in Eq. 1. Specification of the basis assigns units of H which has a unit of bit when digit 2 is taken as the base. Therefore, Eq. 1 can be simplified as Eq.:

$$H = -\sum_{i=1}^{n} p_i \log_2 p_i \tag{2a}$$

Eq. 1 and Eq. 2 have both a negative symbol. In most cases, p_is are less than 1 even though *pi* values range from 0 to 1. The logarithm with base *2* as the, of any values <1 is always negative. Therefore, the value of H was accordingly always negative if there were no negative symbol in either Eq. 1 or Eq. 2a. That is meaningless because uncertainty of a set of outcomes can not be negative. Shannon gave a strictly mathematical proof for Eq. 1 (Shannon, 1948). The symbol has its non-replaceable meaning which ensures the set of outcomes positive because the Shannon entropy can NOT be negative.

In Eq. 2 if all p_is have the same (equal each) value i.e., $p_i = \frac{1}{n}$, then

$$H = -\sum_{i=1}^{n} \frac{1}{n} \log_2 \frac{1}{n} \tag{2b}$$

$$H = -n\frac{1}{n} \log_2 n^{-1} \tag{2c}$$

$$H = n\frac{1}{n}\log_2 n \tag{2d}$$

Therefore, $H = \log_2 n$. H is a monotonically increasing function of n. At this condition, H has its maximum value. On contrary, if every p_i except one is zero, H is minimum and equal to zero. Thus, the actual H may range from zero to $\log_2 n$ (Kawachi et al., 2001). When H reaches the maximum the given set of outcomes have the maximum bits of uncertainty and *vice verse*. Shannon entropy is, therefore, a function estimate of uncertainty associated with probability distribution of a set of random variables in a given system such as NH_3 volatilization from fertilized soils.

Information entropy in NH_3 volatilization from fertilized soils can be quantified based on Shannon's information theory. The probability of the NH_3 volatilization in this study, p_i is defined as

$$p_i = \frac{Q_s}{Q_t} \tag{3}$$

where Q_s is the quantity of volatilized NH_3 from the fertilized soils of a given measurement. Q_t is the total quantity of volatilized NH_3 over a given period of measurement time. The data used in this study are from our previous research which included four soils, four fertilizers and a control without fertilization, three incubation temperatures and two soil water regimes and five measurements: Days 1, 3, 7, 14, and 28 (Liu et al., 2007). Thus,

$$Q_t = \sum_{i=1}^{n} Q_{si} \tag{4}$$

where Q_{si} is the ith Q_s. Shannon entropy of NH_3 volatilization from the fertilized soils is accordingly defined as

$$H_{NH_3} = -\sum_{i=1}^{n} \frac{Q_{si}}{\sum_{i=1}^{n} Q_{si}} \log_2 \frac{Q_{si}}{\sum_{i=1}^{n} Q_{si}} \tag{5}$$

where $n = 5$ because there were five measurements during the incubation period of 28 days (Liu et al., 2007).

Based on Eq. 5, Shannon's information entropy can be used as a measure of soil diversity and homogeneity (Martin and Rey, 2000). Similarly, it can also be used as a measure of homogeneity of NH_3 volatilization from fertilized soils. Figure 1 shows that there are significant ($p = 0.05$) differences in the Shannon entropies of NH_3 volatilization among the four tested soils. In Eq. 5, H_{NH_3}, the Shannon entropy is actually a homogeneity index of

NH_3 volatilization that is regardless with rates or quantity of NH_3 volatilization. For example, at 29°C and 20% FC and with application rate of 75 kg N ha^{-1} of $(NH_4)_2SO_4$, quantity of volatilized NH_3 from the Biscayne Marl Soil (BMS) was 12.02 kg N/ha but that from the Quincy Fine Sand (QFS) was only 1.77 kg N ha^{-1} (Liu et al, 2007). Obviously, N-loss from the the BMS soil was greater than from the QFS soil. However, this does not tell the homogeneity of NH_3 volatilization from the soils. In fact, most (90% of total N-loss) of NH_3 volatilization in the fertilized BMS soil occurred during the first 3 days of 28 days incubation with remaining 10% during the next 4 days. This means that the BMS soil had low holding capacity of NH_4^+ ions from applied fertilizers such as $(NH_4)_2SO_4$ due to its high pH and great electrical conductivity (EC) (Liu et al., 2007). However, in the QFS soil, NH_3 volatilization loss accounted for 49.7%, 14.5%, 17.2% and 18.6% of the total NH_3 volatilization during the 1-3, 4-7, 8-14, and 15- 28 days, respectively. Accordingly, NH_3 volatilization was prolonged in the QFS soil as compared to that in the BMS soil. Therefore, the QFS soil had much smoother NH_3 volatilization from the applied fertilizer than the BMS soil.

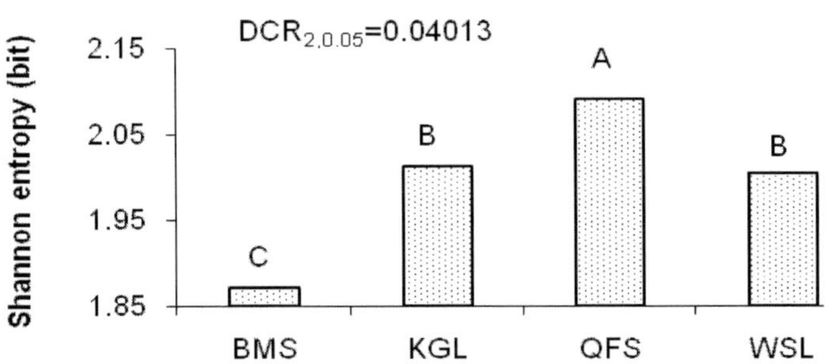

Figure 1. Shannon entropy (bit) of NH3 volatilization for four soils. BMS = Biscayne Marl Soil; KGL = Krome Gravelly Loam; QFS = Quincy Fine Sand; and WSL = Warden Silt Loam. The histograms with different letters differ significantly at p=0.05 based on Duncan's Multiple Test. DCR2, 0.05 = critical range of Duncan's multiple range test between two compared groups at p =0.05.

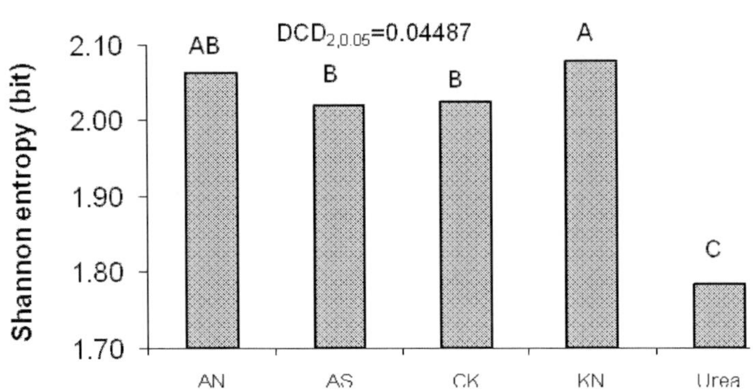

Figure 2. Shannon entropy (bit) of NH3 volatilization for four soils amended with different sources of N fertilizers. AN = ammonium nitrate; AS = ammonium sulfate; CK = control; KN = potassium nitrate,. The histograms with different letters differ significantly at p=0.05 based on Duncan's Multiple Test. DCR2, 0.05 = critical range of Duncan's multiple range test between two compared groups at p =0.05.

Soil with greater H_{NH_3} favors more homogeneous conversion from NH_4^+ to NH_3 and hence more homogeneous NH_3 volatilization as compared to a soil with smaller H_{NH_3}. Therefore, H_{NH_3} may be used as one of the indices to appraise soil quality of smooth and homogeneous NH_3 volatilization from fertilized soils. The crop plants may have more chance to uptake NH_4^+ from the applied N fertilizer due to the smoothness and homogeneity of NH_3 volatilization.

Potassium nitrate (KN) had the greatest Shannon entropy in NH_3 volatilization but urea had the least. Potassium nitrate itself does not contain any NH_4^+ ions. The volatilized NH_3 from the soil amended with KNO_3 might have two possible sources: one part from soil just like the control and the other part from possible reduction of nitrate (NO_3^-) from KNO_3 applied to the soils because cumulative NH_3 volatilization from KN was 5.7-fold greater than that from the control (Liu et al., 2007). The NH_3 volatilization from the tested soils amended with KN might have lasted longer than that from the control due to the possible reduction of NO_3^-. Accordingly, KN had significantly greater Shannon entropy than the control.

Similarly, Shannon entropy was significantly greater for soil amended with AS than that for soil amended with urea. The former N source contains NH_4^+ which can be readily subject to NH_3 volatilization under ideal condition. When urea is used as N source the urea form of N form has to be converted to NH_4^+ form and then to NH_3 form. This may explain, in part, slightly delay in NH_3 volatilization from urea as compared to that from AS. Therefore, NH_3 volatilization from urea follows a mono-peak curve rather than a monotonically decreasing curve as was observed for $(NH_4)_2SO_4$ amended soils. The relative distributions of NH_3 volatilization from the QFS soil amended with urea at 29°C and 80% FC water regimes were 8.8, 46.3, 36.8, 6.2 and 1.9 percent for 1, 2 to 3, 4 to 7, 8 to 14, and 15 to 28 days, respectively. The Shannon entropy values were 1.785 and 2.021 bits for urea and $(NH_4)_2SO_4$, respectively. This indicated that $(NH_4)_2SO_4$ volatilized much more homogeneous than urea.

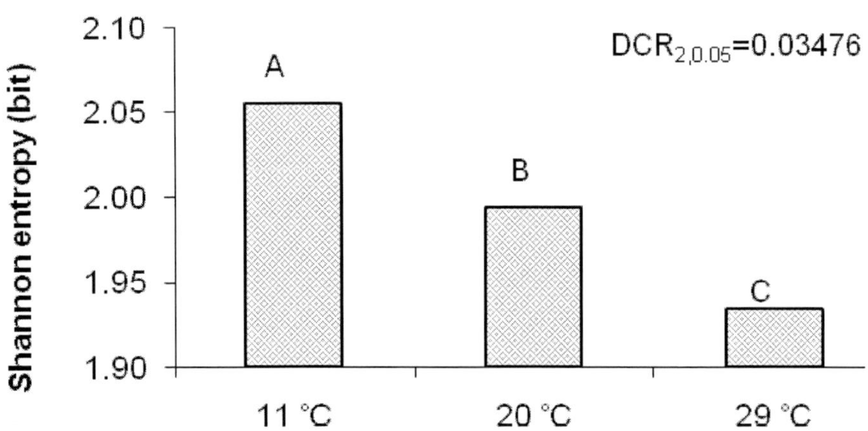

Figure 3. Shannon entropy (bit) of ammonia volatilization across the three temperatures during in cubation of four soils at 20 and 80% field capacity soil water regimes. The histograms with different letters differ significantly at p=0.05 based on Duncan's Multiple Test.

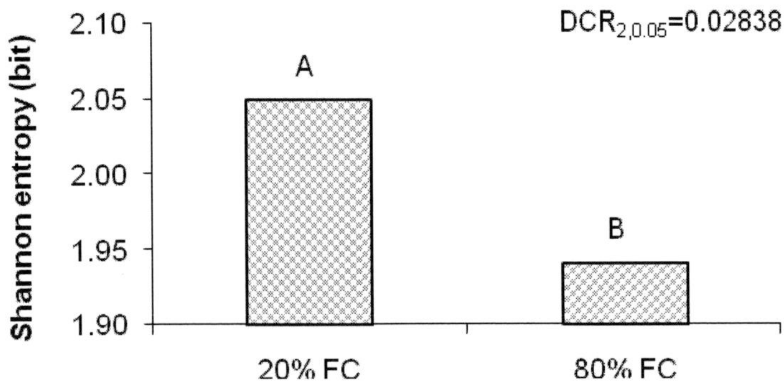

Figure 4. Shannon entropy (bit) of ammonia volatilization at either 20 or 80% field capacity soil water regimes. Data presented as means across four soils, five N sources, three incubation temperatures, and five sampling dates, respectively. The histograms with different letters differ significantly at p=0.05 based on Duncan's Multiple Test.

Shannon entropy of NH_3 volatilization was significantly greater at 20% FC than that at 80% FC soil water regime (Figure 4). Either conversion from NH_4^+ to NH_3 or delivery of NH_3 from aquatic state to gaseous state at 29°C, was much faster than at 11°C (Figure 3). For example, in the QFS soil amended with $(NH_4)_2SO_4$ at 80% FC soil water regime, the distributions of N-loss via volatilization was 43.6, 24.8, 17.2, 8.8 and 5.5 percent for 1, 2 to 3, 4 to 7, 8 to 14, and 15 to 28 days of incubation, respectively, at 29°C while 28.2, 23.3, 26.3, 14.5, and 7.7, respectively, at 11°C (Liu et al., 2007). Accordingly, the rates of NH_3 volatilization were significantly more homogenous at 11°C than those at 29°C.

The Shannon entropy of NH_3 volatilization was significantly greater at 20% FC than that at 80% FC soil water regime, indicating that NH_3 volatilization was significantly homogenous at the former than that at the latter soil water regime. Indeed, the relative distributions of total N-loss via volatilization from the QFS soil amended with $(NH_4)_2SO_4$ at 20°C were 20.9, 19.8, 20.6, 19.0, and 19.7 percent for 1, 2 to 3, 4 to 7, 8 to 14, and 15 to 28 days, respectively at 20% FC soil water regime, while 32.0, 31.6, 15.1, 13.8, and 7.5percent, respectively, at 80% FC soil water regime was (Liu et al. 2007). At 20% soil water regime, there was not much chemisorption and physical-sorption for the applied NH_4^+ ions in the QFS soil due to extreme deficit soil water. This could explain, in part, greater NH_3 volatilization at 20% than that at 80% FC soil water regime.

Using the Shannon entropy values as an index of homogeneity of NH_3 volatilization the soils used in this study can be categorized into the following patterns: (1) Homogeneous Volatilization Pattern such as the QFS soil. This pattern has a Shannon entropy value of $\geq$ 2,1 bits, the NH_3 volatilization occurred smooth and relatively slowly from the beginning to the end during the incubation period; (2) Semi-Homogeneous Volatilization Pattern including the KGL and WSL soils. This pattern has a Shannon entropy value of >1.9 but <2.1 bits and less smooth NH_3 volatilization than Pattern 1;and (3) Non-Homogeneity Volatilization Pattern: the BMS soil (Figure 1) is an example of this pattern. This pattern has a Shannon entropy value of $\leq$ 1.9 bits. It has a high rate of NH_3 volatilization for a short period soon after application of N fertilizer or 90% N-loss via volatilization happens in the first three days during the incubation period. Similarly, the three fertilizers containing either NH_4^+ source or other N forms that can be converted to NH_4^+ form can be divided into two types: (1)

Homogeneous Volatilization Fertilizers: the Shannon entropy value is <1.9 bits with relative homogeneous volatilization through the incubation period. Both NH_4NO_3 and $(NH_4)_2SO_4$ belong to this type (Figure 2); and (2) Non-Homogeneity Volatilization Fertilizers: their Shannon entropy value is >1.9 bits with heterogeneous volatilization. Urea is classified as this type (Figure 2). Likewise, (1) 11°C is a Homogeneous Volatilization Temperature; (2) 29°C is a Non-Homogeneous Volatilization Temperature; and (3) 20°C is a Semi-Homogeneous Volatilization Temperature (Figure 3). Twenty percent FC is a Homogeneous Volatilization Soil Water Regime that has a steady and similar volatilization rate over the entire incubation period but 80% FC a Non-Homogeneous Volatilization Soil Water Regime (Figure 4).

The homogeneity index of NH_3 volatilization and classifications of the soil, fertilizers, incubation temperatures, and soil water regimes based on the index do describe the homogeneity status of NH_3 volatilization during the incubation period . However, the Shannon rate of NH_3 volatilization can tell the environmental friendliness of the fertilizers or soils because the Shannon rate presents the quantity of volatilized NH_3 per unit bit per unit area to the environment from fertilized soils.

Shannon entropy has a close relationship with soil particle size distribution (Figure 5). The sandy particles positively associate with the Shannon entropy. The more the sandy particles distribute, the greater the Shannon entropy is and *vice versa*. The more sandy the soil is the homogeneity of NH_3 volatilization is greater for sandy soils as compared to that for fine texture soils. This may be associated with NH_4^+ chemisorptions and physical-sorption and their corresponding adsorptions, enzymatic reactions, and N mineralization of the applied fertilizers to soils but more research is needed.

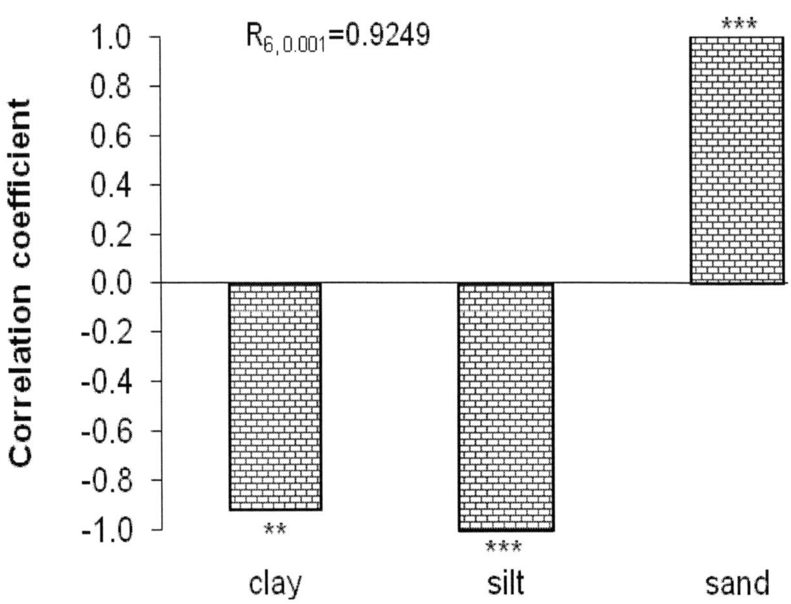

Figure 5. Relationship between soil particle size distribution and Shannon entropy of ammonia volatilization. The correlation coefficient is significant at either 0.01 (**) or 0.001 (***). R = -0.9171, -0.9961, and 0.9995 for clay, silt, and sand, respectively. R6, 0.001 = 0.9294 means the critical value of the correlation coefficient between Shannon entropy and each of the three kinds of soil particles at degree of freedom of 6 and p = 0.001.

SHANNON RATE OF AMMONIA VOLATILIZATION

Based on the Shannon entropy, we can further define the Shannon rate (R_s) of NH_3 volatilization as follows:

$$R_s(gN \cdot ha^{-1} \cdot bit^{-1}) = \frac{\sum_{i=1}^{n} Q_{si}}{H_{NH_3} \times A} \tag{6}$$

where R_s = the Shannon rate of NH_3 volatilization ; A = area (hectare). Please see Eq. 5. for explanation of other variables.

Under the same environmental condition, Shannon rate of NH_3 volatilization can differ significantly (Figure 6) in the. The R_s values were nearly 2-fold greater for the BMS and KGL soils as compared to that for the QFS and WSL soils. This may be associated with soil pH and salt saturation, as evidence from greater pH and electrical conductivity (EC) for the former soils than those for the latter soils (Liu et al., 2007). Soils which favor greater NH_3 volatilization can be considered as less environmentally sustainable as compared to those which favor lower NH_3 volatilization. The trend in R_s values across different soils reflects the trend in cumulative NH_3 volatilization from the soils (Liu et al. 2007).

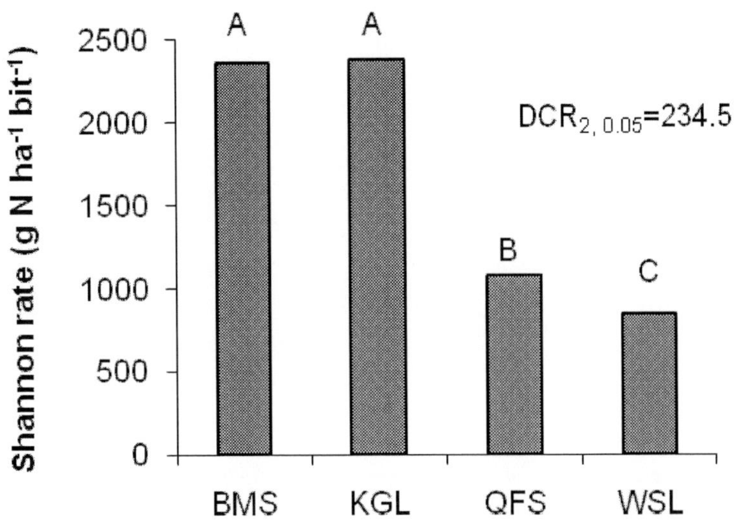

Figure 6. Shannon rate (g N ha-1bit-1) of ammonia volatilization cross the four tested soils. BMS denotes Biscayne Marl Soil KGL Krome Gravelly Loam, QFS Quincy Fine Sand, and WSL Warden Silt Loam. The histograms with different letters differ significantly at p=0.05 based on Duncan's Multiple Test.

The Shannon rates of different N fertilizers decreased in the order: urea > AS > (Figure 7) which is the order of cumulative NH_3 volatilization from the fertilizers. There was no significant difference in R_s values across three temperatures evaluated in this study (Figure 8). The R_s of NH3 volatilization was 3-fold greater at 20% FC that that at 80% FC soil water regime (Figure 9).

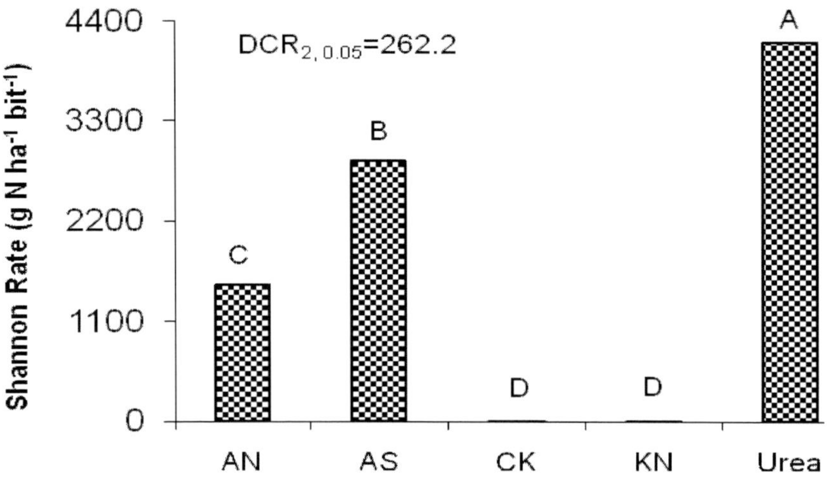

Figure 7. Shannon rate (g N ha-1bit-1) of ammonia volatilization for different N sources and for soil with no N amendment. AN = ammonium nitrate; CK = control; AS = ammonium sulfate; KN = potassium nitrate. The histograms with different letters differ significantly at p=0.05 based on Duncan's Multiple Test.

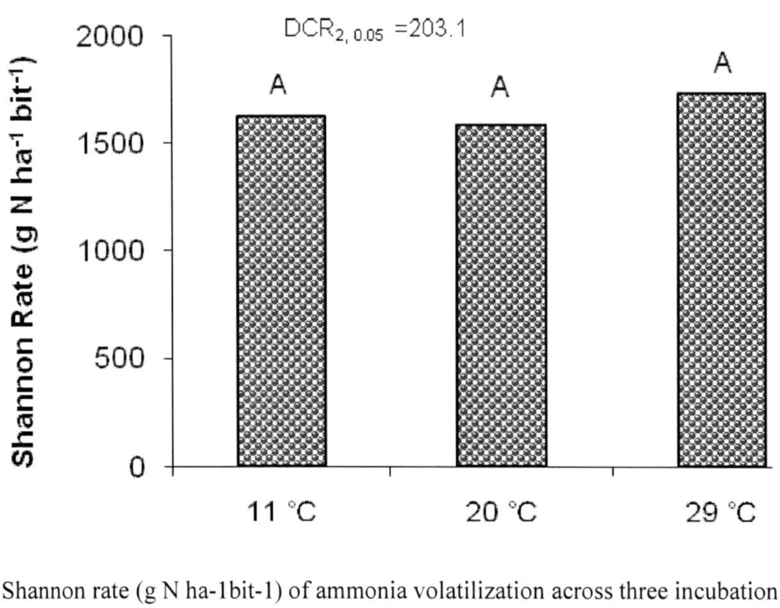

Figure 8. Shannon rate (g N ha-1bit-1) of ammonia volatilization across three incubation temperatures. The histograms with different letters differ significantly at p=0.05 based on Duncan's Multiple Test.

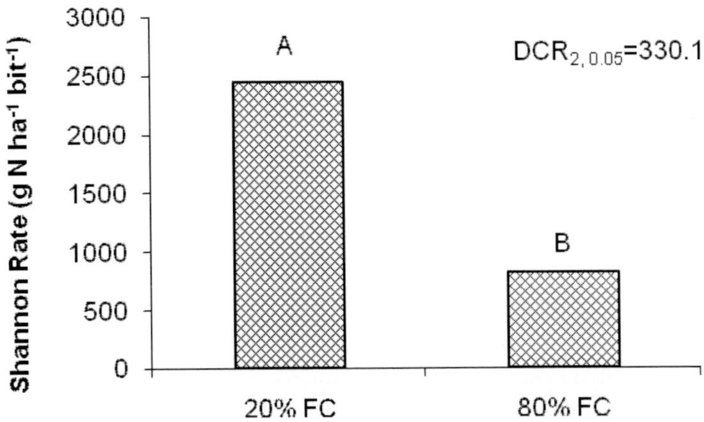

Figure 9. Shannon rate (g N ha-1bit-1) of ammonia volatilization at two soil water regimes. FC means field capacity. The histograms with different letters differ significantly at p=0.05 based on Duncan's Multiple Test.

CONCLUSIONS

Shannon's information theory has been applied to multidiscipline such as ecology, hydrology and molecular biology. We have attempted to use this concept to describe NH_3 volatilization from four different soils which received four different N sources incubated at 3 temperatures and 2 soil water regimes. Shannon entropy of NH_3 volatilization represents homogeneity index of NH_3 volatilization from different soils amended with various N fertilizers. Using the Shannon entropy data, Shannon rate values were calculated for NH_3 volatilization, which mimics the cumulative NH_3 volatilization a given set of conditions for. This study resulted in the following conclusions:

1) The pattern of NH_3 volatilization during 28 days incubation was: (i) homogeneous for the QFS soil; (ii) non-homogeneous for the BMS soil; and (iii) semi-homogeneous for the KGL and WSL soils,
2) Ammonia volatilization pattern was (i) homogeneous for NH_4NO_3 and $(NH_4)_2SO_4$; (ii)non-homogeneous for urea,
3) Among the incubation temperatures, NH_3 volatilization pattern was (i) homogeneous at 11°C; (ii) semi-homogeneous at 20°C; and (iii) non-homogeneous at 29°C.
4) Likewise, NH_3 volatilization pattern was homogeneous at 20% FC, and non-homogeneous at 80% FC soil water regime.
5) The Shannon rates of NH_3 volatilization for the N fertilizers containing ammonium or potential ammonium followed the order of $NH_4NO_3 < (NH_4)_2SO_4 <$ urea. Similarly, the four soils obeyed the order of WSL < QFS < BMS < KGL. The soil water regimes complied with the order 80%<20%.
6) More research is needed for the applications of these new concepts in both agriculture and environment sciences.

ACKNOWLEDGEMENT

This research was supported by the Florida Agricultural Experimental Station and a grant from USDA-ARS.

REFERENCES

Abe, S. 1997. A note on the *q*-deformation-theoretic aspect of the generalized entropies in nonextensive physics. *Physics Letters A.* 224: 326-330.

Asman, W. A. H.; Jonker, P. J.; Slanina, J.; Baard, J. H. 1982. Neutralization of acid in precipitation and some results of sequential rain sampling. In: *Deposition of Atmospheric pollutants*. Georgii, H. W.; Pankarth, J. (eds.). D. Reidel Publishing Co., Dordrecht, The Netherlands. Pp. 115-124.

Aneja, V. P.; Murray, G. C.; Southerland, J. 1998. Atmospheric Nitrogen Compounds: Emissions, Transport, Transformation, Deposition and assessment. *Environmental Manager.* (April Issue), 22-25.

Bonchev, D. And D. H. Rouvray. 2005. Complexity in Chemistry, Biology, and Ecology. Pp. 1-344. Springer Science+Business Media, Inc. 233 Spring Street, New York, NY 10013, USA.

Brunekreef, B.; Holgate, S. T. 2002. Air pollution and health. *Lancet*. 360: 1233-1242.

Chakrabarti, C. G. and I. Chakrabarty. 2005. Shannon entropy: amiomatic characterization and application. *International Journal of Mathematics and Mathematical Science.* 17: 2847-2854.

Dentener, F. J.; Crutzen, P. J. 1994. A three-dimensional model of the global ammonia cycle. *J Atmospheric Chemistry*. 19: 331-369.

Duce, A. R.; Liss, P. S.; Merrill, J. T.; Atlas, E. S.; Buat-Menard, P.; Hicks, B. B.; Miller, J. M.; Prospero, J. M.; Arimoto, R.; Church, T. M.; Ellis, W.; Galloway, J. N.; Hansen, L.; Jidkells, T. D.; Knap, A. H.; Reinhardt, K. T.; Schneider, B.; Soudin, A.; Tokos, J. J.; Tsunogai, S.; Wollallas R.; Zhou, M. 1991. The atmospheric input of trace species to the world ocean. Global *Biogeochemical Cycles*. 5: 193-259.

FAO/IFA. 2001. Global Estimates of Gaseous Emissions of NH3, NO and N_2O from agricultural land. Food and Agriculture Organization of the United Nations (FAO) / International Fertilizer Industry Association (IFA), Rome, 106. (Available from http://www.fertilizer.org/ifa).

Fenn, L. B.; Hossner, L. R. 1985. Ammonia Volatilization from Ammonia or Ammonium-Forming Nitrogen Fertilizers. In *Advances in Soil Sciences*. Stewart, B. A. (eds.) Springer-Verlag, New York, USA. pp 123-169.

Ferm, M. 1998, Atmospheric ammonia and ammonium transport in Europe and critical loads: a review. *Nutrient Cycling in Agroecosystem*. 51: 5-17.

Finlayson-Pitts, B. J.; Pitts, J. N. Jr. 1986. *Atmospheric Chemistry: Fundamentals and experimental Techniques.* Wiley: New York, NY.

Finlayson-Pitts, B. J.; Pitts, J. N. Jr. 2000. *Chemistry of the Upper and Lower Atmosphere.* Academic Press. San Diego, CA. pp. 265-293; 763-843.

Fleming, S .W. 2007. An information theoretic perspective on mesoscale seasonal variations in ground-level ozone. *Atmospheric Environment*. 41: 5746-5755.

Frederick, K. K., M. S. Marlow, K. G. Valentine, and A. J. Wand. 2007. Conformational entropy in molecular recognition by proteins. *Nature*. 448: 325-329.

Hill, M. O. 1973. Diversity and Evenness: A Unifying Notation and Its Consequences *Ecology*. 54: 427-432.

Kawachi, T. T. Maruyama, and V. P. Singh. 2001. Rainfall entropy for delineation of water resources zones in Japan. *Journal of Hydrology*. 246: 36-44.

Liu, G. D.; Li, Y. C.; Alva, A. K. 2007. High water regime can reduce ammonia volatilization from soils under potato production. *Commun. Soil Sci. Plant Anal*. 38: 1203-1220.

Marcazzan, G. M.; Vaccaro, S.; Valli, G.; Vecchi, R. 2001. Characterization of PM_{10} and $PM_{2.5}$ particulate matter in the ambient air of Milan (Italy). *Atmosphere Environment*. 35: 4639-4650.

Marenduzzo, D., C. Micheletti, and P. R. Cook. 2006. Entropy-driven genome organization. *Biophysical Journal*. 90: 3712-3721.

Martín, M. A. and J.M. Rey. 2000. On the role of Shannon's entropy as a measure of heterogeneity. *Geoderma*. 98: 1-3

Shannon, C.E. 1948 A Mathematical Theory of Communication. *Bell System Technical Journal*. 27: 379-423; 623-656.

Shannon, C.E. and Weaver, W. 1949. The Mathematical Theory of Communication. University of Illinois Press, Urbana Reprinted in 1963 and 1998.

Schlesinger, W. H.; Hartley, A. E. 1992. A global budget for atmospheric NH_3. *Biogeochemistry*. 15: 191-211.

Van Breeman, N.; Burrough, P. A.; Velthorst, E. J.; Van Dobben, H. F.; De Wit, T.; Ridder, T. B.; Reijnders, H. F. R. 1982. Soil acidification from atmospheric sulphate in forest canopy through-fall. *Nature*. 299: 548-550.

Warneck, P. 1999. *Chemistry of the Natural Atmosphere*. Academic Press, London.

In: Agricultural Systems: Economics, Technology and Diversity ISBN 978-1-60692-025-1
Editor: Oliver W. Castalonge

Chapter 4

ECONOMIC ANALYSIS OF PRECISION FARMING TECHNOLOGIES AT THE FARM LEVEL: TWO GERMAN CASE STUDIES

Andreas Meyer-Aurich[*], Markus Gandorfer[**] and Alois Heißenhuber[**]

[*]Leibniz-Institute for Agricultural Engineering Potsdam-Bornim (ATB) Potsdam, Germany

[**]Institute of Agricultural Economics and Farm Management Technische Universität München, Freising-Weihenstephan, Germany

ABSTRACT

The aim of this chapter is to evaluate the economic whole farm impact of the implementation of precision-farming technologies. Therefore, two farms were analyzed, which had implemented different precision agriculture technologies several years ago. The farms can be regarded as early adopters of precision agriculture technologies in two very different agricultural production areas. One of the two analyzed farms (1560 ha) is located in East Germany (Saxony-Anhalt), in a region with low annual rainfall and high fertile soils. In this farm nitrogen and potassium fertilizer applications, fungicides applications, and seeding are managed site-specifically following a mapping-approach, based on yield maps and aerial photographs. The other farm (150 ha) is located in Bavaria, in a region with high annual rainfall and comparatively fertile soils. In this farm the Yara-N-Sensor® is used for site-specific nitrogen management. The economic analysis is based on economic efficiency calculations for the implemented technologies, e.g., break-even-analysis and profit estimations. Furthermore, the two farmers were interviewed about the impact on their farm management and organization.

The implementation of the technologies provided economic gains for both of the farms. While the implementation of the sensor-based technology had only moderate effects on the overall farm management, the implementation of the mapping based system resulted in a far-reaching reorganization of the farm.

We conclude that a sensor-approach is easier to implement than a mapping approach. The implementation of a mapping-based site-specific management approach requires more skills, e.g., for analyzing site-specific data or creating application maps. Positive returns on the investments in the fertilizer technology can only be achieved when the technologies are used on sufficient acreage. This is easier to realize on big farms in eastern Germany or in terms of a collaborative use of the technologies than on comparatively small farms in southern Germany. However, in both regions we didn't find evidence that the innovation is adopted by neighboring farmers, which indicates that there are still considerable adoption constraints that limit the diffusion of precision farming technologies in Germany.

Key words: economics of precision farming, sensor-approach, mapping-approach

INTRODUCTION

Despite the great interest in precision farming technologies from researchers and the public, the adoption of many of these technologies by farmers in Germany and elsewhere is slow [5, 8]. The identified reasons for low adoption rates are high costs of adoption, uncertainty in returns due to adoption, and lack of demonstrated effects of the technologies on yields and input use [10]. Given the high costs of most of the technologies, the economic profitability can be identified as a major determinant for the adoption.

The concept of Precision Farming can be understood as the use of information and technology to improve farming operations. Often the understanding of the concept is limited to variable rate technologies, but the technological possibilities cover a great range from automatic data acquisition over site-specific fertilization to optimized fleet management [2]. The challenge of Precision Farming is that different computer based technologies need to be implemented to perform information based farming efficiently. Swinton and Lowenberg-DeBoer (1998) identified four different component sets of computer-based technologies, which are needed for site-specific farming. These are Geographic Information Systems (GIS), Global Positioning Systems (GPS), controller or variable rate technologies and sensing technologies. However, for the different Precision Farming technologies not necessarily all components have to be used. It is obvious that with the number of information technologies involved the complexity of the management systems increases and requires specific skills by the users and infrastructure for electronic communication. This may be an adoption constraint for more sophisticated approaches of Precision Farming.

Many economic studies of Precision Farming are restricted to the effects the implementation of variable rate technology and have provided mixed results [1, 4, 7, 11, 12, 16, 17]. It has been shown that sophisticated models are necessary to prove an economic effect of the implementation of a precision agriculture technology due to the necessity of modeling spatial effects [1, 6], which contributes uncertainty to economic assessments of Precision Farming implementations. Furthermore, Liu et al. (2006) showed that economic profitability of site-specific management may not be consistent over time. Pannell (2005) argues that because of flat profit functions site-specific variation of input levels can only contribute a very limited extent to higher benefits. However, in order to draw an unbiased picture of the impact of precision farming technologies, it has been stated that the impact assessment of precision agriculture technologies should focus on the whole farm rather than only experimental fields [13]. To date only few studies have analyzed the whole farm impact of precision farming technologies. Batte and Arnholt (2003) analyzed the benefits from Precision Farming for six early adopters in Ohio (USA) and concluded that all farmers derived value from their concept of Precision Farming. However, the farmers appeared to value information gathering technologies higher than variable rate technologies [3]. Other studies have analyzed factors which lead to the adoption of Precision Farming [9, 15]. They found farm size, human capital, and innovativeness of the farmer and ownership of the land as predictors for the adoption of Precision Farming technologies. However, adoption rates only indirectly provide evidence of economic benefits of these technologies.

The aim of this contribution is to investigate the whole farm impacts of the implementation of Precision Farming technologies in order to identify the specific benefits of the technologies for two farms in Germany, which can be considered as early adopters of Precision Farming technologies. We provide a conceptual framework which can also be used to assess whole farm impacts of other precision agriculture technologies.

Conceptual Framework

The whole farm impact assessment of Precision Farming technologies should take into account all aspects of farm management, including effects on crop yield, input use, changes in management, and the quality of work. In general Precision Farming can be considered as information driven farming, where advanced information results in improved management. However, information and information processing may be cost intensive, which must be covered by the benefits from the use of information to be attractive for farmers. The sources of costs of Precision Farming and possible benefits are listed in Table 1. We distinguish four different cost types, which are associated with the implementation of Precision Farming technologies. Information costs come from the necessary investments in technologies or rental fees for sensing specific information. Costs for data processing and the adapted management have to take into account the information, i.e., with specific software or hardware products, which adjust the management according to the gathered information. Further, it has to be expected that learning costs apply. Learning costs can be additional time for the development of management schemes, calibration of the machinery or opportunity costs due to inefficient use of the Precision Farming technology.

Table 1. Costs and possible benefits from precision farming

Costs of Precision Farming	Possible Benefits
Information costs	Crop yield effects
Costs of data processing	Changes in input use
Costs of adapted management	Changes in Management
Learning costs	Quality of work

Possible benefits from Precision Farming stem from crop yield effects, changes in input use, the management, and the quality of work. Precision Farming can result in higher crop yields with reduced inputs or vice versa. The implementation of Precision Farming has consequences in the labor requirements and the skills of the employed workers.

The profitability of the Precision Farming approach depends on the cost assessment and the estimation of the response of the agricultural system to the advanced information management. The costs depend on the price of the technology, opportunity costs for labor and capital, and the useful life of the investment. While the costs can be estimated rather precisely the response of the system to the improved management can only be assessed on the basis of assumptions, since there are no data available of a reference system.

Two Case Studies

In order to identify benefits and constraints of the adoption, the impact of the adoption of Precision Farming technologies has been investigated on two farms in Germany, which can be considered as early adopters of Precision Farming technologies. The farms have implemented different approaches to Precision Farming in very different environmental settings. Besides the analysis of the economic effects our investigations laid emphasis on non-economic effects of the adoption in order to identify important criteria for a wider diffusion of the technologies. Therefore, the potential of diffusion from both of the farms was estimated.

Case Study Farm A

Farm A operates on about 1500 ha of high fertile soil in Saxony-Anhalt with rain fed crop production. Since 1998 the farm has implemented Precision Farming technologies to realize a sophisticated concept of zone management (mapping approach). Yearly precipitation is low with 450 mm per year, which regularly results in water deficits in crop production. Besides crop production the farm produces milk and beef with a herd of 350 cows. However, precision agriculture technologies are only implemented in crop production. The motivation of the farmer to start with Precision Farming came from the realization of the great spatial heterogeneity of his fields while he completed his pilot license. Since the farmer has a strong interest in crop production, he was convinced that the heterogeneity of thc plant growth should result in an adjusted management. Therefore, he started to investigate the sources of the heterogeneity on the fields of his farm. He participated in research projects, which analyzed spatial yield patterns on his fields. On some of his fields he analyzed the soil for the plant nutrients, i. e. P and K within some of his plots on a one ha raster basis. The farmer found that nutrient levels were high in parts of the fields where yields were low and on sites with higher yields nutrient levels were low. Thus, the farmer reduced the application of P and K fertilizers on parts of the fields with high levels of nutrients and low yield levels. In the following years he analyzed the nutrient content of the soil on some selected sites to see how the soil responded to reduced fertilization. Based on the combined yield mapping he got the additional information of the crop response, which gave him a better understanding of the whole system. Later he extended the fertilization practice to the nitrogen management, seeding rates, and fungicide application. The whole Precision Farming management is based on a mapping approach, where previous yields and soil nutrients determine the patterns of the management. However, the farmer adjusts the level of fertilizer rates according to actual and expected weather conditions, even though this does not affect the distribution patterns of the applied fertilizers.

Costs and Benefits of the Precision Farming Approach in Farm A

With the implementation of the Precision Farming approach, the farmer has invested working time and specific machineries. Costs were estimated with an interest rate of 4% and a useful life of the investment of ten years. Additionally to the machine costs we calculated opportunity costs for the development of the Precision Farming concept. We calculated the benefits of the implementation primarily on the basis of the reduced inputs into the system. This was compared to the situation before the farmer had implemented the Precision Farming technologies. Further we assumed an increase in crop yield of 5% due to site-specific variation of the seeding rates according to the experiences of the farmer. With the new management concept the farmer saves the costs of a manager who coordinated the work on the farm before. Instead the farmer himself now coordinates the work of the workers with the help of the application maps. Table 2 shows costs and assumptions for the cost-benefit analysis, which results in a yearly profit increase of 49 € ha^{-1}. Besides the economic benefits the farmer states that the introduction of Precision Farming technologies on his farm has contributed to additional non-financial benefits. The field workers on the farm are all involved in the variable application technology and very much motivated to improve the cropping system. The communication between the farmer and his workers has been facilitated. This is due to the fact each worker now gets an application map in the morning with all operating instructions. He no longer needs to explain to the workers where they should go and which specialties they should consider.

Table 2. Cost-benefit analysis for Farm A

Yield mapping equipment (Agromap Basic)	16,000	€
Meteorological station	3,000	€
Computer equipment	3,000	€
Aerial pictures	3,000	€
Additional equipment for variable fertilizer application	5,000	€
Additional equipment for variable seeding	5,000	€
Equipment for site-specific plant protection	10,000	€
2 ACT Terminals	10,000	€
Soil sampling	6,000	€
Opportunity costs for the development of the Precision Farming concept	22,500	€
$\sum$ initial cost	83,500	€
Annual cost	10,244	€ a^{-1}
Annual cost for mounting, repairing and updates (2% of the initial cost)	1,670	€ a^{-1}
Annual labor cost to process the application maps (0,5 months)	1,250	€ a^{-1}

Table 2. (Continued)

∑ annual cost	13,164	€ a^{-1}
Cropping area	1,500	ha
Annual cost per unit area	9	€ a^{-1} ha^{-1}
Reduced fertilizer cost	23,800	€ a^{-1}
Reduced seed cost	5,600	€ a^{-1}
Reduced fungicide cost	13,100	€ a^{-1}
Reduced operating cost	30,000	€ a^{-1}
Higher yields due to site-specific seeding	13,500	€ a^{-1}
∑ higher yield and reduced input cost	86,000	€ a^{-1}
	57	€ a^{-1} ha^{-1}
Profit increase	72,836	€ a^{-1}
	49	€ a^{-1} ha^{-1}
Assumptions		
Useful life	10	years
Interest rate	4	%
Annual labor cost	30,000	€ a^{-1}

Case Study Farm B

Farm B operates on 150 ha in Bavaria. Average annual rainfall for the region is 750 mm with rainfall distribution approximately uniform over the year. This results in moderate to high crop yields (wheat: 8 Mg ha^{-1}) with less production risk than in the case of farm A. Instead the farmer sometimes faces uncontrolled mineralization of soil nutrients (nitrogen), which result in lodging and yield penalties. Thus the farmer is sensitive to over-fertilization of his fields and is interested to adjust the amount of fertilizers, i.e., nitrogen to the needs of the plants. He reported that he has always been interested in adjusting the application rates to the nutrient status of the plants and did that in former times "by eye". With the emergence of sensor-based technologies to analyze the nutrient status of the plants during application he was very interested in experimenting with the new technologies on his fields. He participated in a research project from 1999 to 2001, which subsidized the investment costs of the "Yara-N-Sensor®" by 50%. Within the research project he also experimented with mapping based concepts to adjust the application of inputs. However, he was not convinced by the results of the mapping based Precision Farming concepts. Therefore, he still uses only the online-sensor approach for the application of nitrogen for the second and third application to wheat and rye.

Costs and Benefits of the Precision Farming Approach in Farm B

The cost of the Precision Farming approach in farm B stems basically from the online-sensor, which is commercially available (Yara-N-Sensor®). Furthermore, additional labor input for mounting and calibration of the sensor has to be accounted for. As stated earlier, the response of the agricultural system to the modified management can only be estimated on the basis of assumptions and assessments of the farmer.

Table 3. Cost-benefit analysis for Farm B

	Without investment subsidy	With investment subsidy	Unit
Yara N-Sensor (annual cost)	2,699	1,104	€ a^{-1}
Annual cost for mounting, repairing and updates 2% of the initial cost	440	440	€ a^{-1}
∑ annual cost	3,139	1,544	€ a^{-1}
Cropping area	80	80	ha
Annual cost per unit area	39	19	€ $ha^{-1}a^{-1}$
Required yield increase for break-even	0.2	0.1	Mg ha^{-1}
	2.5	1.2	%
Required reduction of N-input use for break-even	52	26	kg N ha^{-1}
	25	12	%
Yield increase with site-specific management (farmer's estimation)	3.5	3.5	%
	0.28	0.28	Mg ha^{-1}
	56	56	€ ha^{-1}
Fertilizer reduction with site-specific management (farmer's estimation)	5	5	%
	11	11	kg ha^{-1}
	8	8	€ ha^{-1}
∑ higher yield and reduced N fertilizer cost	64	64	€ $ha^{-1}a^{-1}$
Change in net benefit	25	45	€ $ha^{-1}a^{-1}$

Assumptions
Initial cost Yara N-Sensor: 22,000 €
Useful life: 10 years
Interest rate: 4%
Price of winter wheat: 200 € Mg^{-1}
Price of nitrogen fertilizer: 0.75 € kg^{-1}
Average uniform nitrogen rate (N): 210 kg ha^{-1}
Average yield expectation (winter wheat): 8 Mg ha^{-1}

Based on the experiences of the farmer he estimates a 3.5% increase in crop yields and a 5% reduction of fertilizer inputs due to the sensor technology. Furthermore, the farmer reported that in some years he expects to harvest higher and more homogenous grain quality. Overall, we found a yearly increase in profit of 45 € ha^{-1}. However, this takes into account the investment subsidy which was paid by the research project, the farmer was involved in. Without subsidy we can find a yearly increase in profit of 25 € ha^{-1} under the stated assumptions.

Conclusion

The Precision Farming approaches of the two analyzed farms differ considerably in the cost structure and the skills and requirements which are needed to follow the respective

Precision Farming approach. The mapping approach, followed by farm A, requires the analysis of different geo-referenced data on yield, soil nutrient status, and aerial pictures to generate application maps and enable the variable application of inputs. It should be noted, that the management at the sub-field level increases the decision possibilities for the farmer and thus the complexity of the management system. In contrast, the N-Sensor requires fewer decisions on the management and doesn't need the processing of geo-referenced information. However, the results of our interviews suggest that both farmers feel comfortable with the system they use and are not convinced of the other system. This implies that the advanced use of information with Precision Farming technologies doesn't necessarily mean that more information must be processed and more decisions must be made by the farmers. Instead, it can be assumed that the majority of farmers are rather interested in reducing the complexity of the already very complex cropping system than increasing it by making the operating decisions at the subfield level.

The profitability analysis of the two farms showed that in both environmental settings the implementation of the Precision Farming concept resulted in higher net benefits and additional non-economic advantages. However, the economic analysis is based on subjective assessment of the response of the cropping system to the implementation of Precision Farming technologies. It can be questioned, if the stated increases in yields and reductions of inputs can only be achieved with the investment in the Precision Farming technology. For example, farm A has achieved a substantial economic benefit by reducing nutrient inputs into the system. It still has to be proved that the site-specific reduction of inputs has an economic advantage over the uniform reduction of inputs without Precision Farming technologies.

The little interest of neighboring farmers in the Precision Farming technologies in both regions indicate that farmers are still skeptical that the implementation of Precision Farming technologies is beneficial for them. In both analyzed farms the interest in the spatial heterogeneity of the fields and the wish to manage this heterogeneity dominated financial considerations. Therefore, we conclude that this interest is crucial for an adoption of Precision Farming technologies, at least in the field of site-specific management.

References

Anselin, L., Bongiovanni, R. and Lowenberg-DeBoer, J. (2006). A Spatial Econometric Approach to the economics of site-specific nitrogen Management in Corn Production. *American Journal of Agricultural Economics* 86, 675-687.

Auernhammer, H. (2001). Precision farming – the environmental challenge. Computers and Electronics in Agriculture 30, 31-43.

Batte, M.T. and Arnholt, M.W. (2003). Precision farming adoption and use in Ohio: case studies of six leading-edge adopters. *Computers and Electronics in Agriculture* 38, 125-139.

Gandorfer, M., Meyer-Aurich, A. and Heißenhuber, A. (2004). Der Einsatz moderner Informationstechnologien im Pflanzenbau – ein Beispiel aus der teilflächenspezifischen Stickstoffdüngung. *Zeitschrift für Agrarinformatik* 2/04: 27-33.

Griffin, T. W., Lowenberg-DeBoer, J., Lambert, D. M., Peone, J., Payne, T.,and Daberkow, S. G. (2004). *Adoption, Profitability, and Making Better Use of Precision Farming Data.* Staff Paper #04–06, (Department of Agricultural Economics, Purdue University, USA).

Hurley, T.M., Malzer, G.L. and Kilian, B. (2004). Estimating Site-Specific Nitrogen Crop Response functions: A conceptual framework and Geostatistical Model. *Agronomy Journal* 96,1331–1343.

Hurley, T.M., Oishi, K. and Malzer, G.L. (2005). Estimating the Potential Value of variable rate Nitrogen applications: A comparison of spatial econometric and geostatistical models. *Journal of Agricultural and Resource Economics* 30, 231-249.

Jürgens, C. (2006). Langsames, aber stetiges Wachstum. Beobachtungen zur Akzeptanz von Precision Farming in Deutschland. *Neue Landwirtschaft* 1/2006,45–47.

Khanna, M. (2001). Sequential adoption of site-specific technologies and its implications for nitrogen productivity: A double selectivity model. *American Journal of Agricultural Economics* 83, 35-51.

Khanna, M., Epouhe, O., F.and Hornbaker, R. (1999). Site-Specific Crop Management: Adoption Patterns and Incentives. *Review of Agricultural Economics* 21, 455–472.

Lambert, D. M. and Lowenberg-DeBoer, J. (2003). Economic Analysis of Row Spacing for Corn and Soybean. *Agronomy Journal*, 95, 564-573.

Liu, Y., Swinton, S.M. and Miller, N.R., (2006). Is site-specific yield response consistent over time? Does it pay? *American Journal of Agricultural Economics* 88, 471–483.

McBratney, A., Whelan, B., Ancev, T and Bouma, J. (2005). Future Directions of Precision Agriculture. *Precision Agriculture* 6, 7–23.

Pannell, D.J. (2006). Flat-earth economics: The far-reaching consequences of flat payoff functions in economic decision making. *Review of Agricultural Economics* 28: 553-566.

Roberts, R.K., English, B.C. and Larson, J.A. (2002). Factors affecting the location of Precision Farming Technology adoption in Tennessee. Journal of Extension (Online) 40 (1). Available at http://joe.org/joe/2002february/rb3.html.

Swinton, S.M. and Lowenberg-DeBoer, J. (1998). Evaluating the Profitability of Site Specific Farming. *Journal of Production Agriculture* 11, 439-446.

Thrikawala, S., Weersink, A., Kachanoski, G. and Fox, G. (1999). Economic Feasibility of variable-rate technology for nitrogen on corn. *American Journal of Agricultural Economics* 81, 914-927.

In: Agricultural Systems: Economics, Technology and Diversity ISBN 978-1-60692-025-1
Editor: Oliver W. Castalonge

Chapter 5

RESTRUCTURING LIVESTOCK FARMS UNDER OIL BOOM ECONOMY IN TRANSITION COUNTRY: CASE STUDY OF MANGISTAU OBLAST, KAZAKHSTAN

Wataru Yamamoto[1]*, Leonid Gurevich[2] and Aigul Kabinova[3]

[1]RECS International Inc., 7-20-11 Nishishinjuku Shinkuku-ku
Tokyo, Japan 160-0023
[2]BISAM Central Asia, Office 236, 221 Bogenbay St. 050026,
Almaty Kazakhstan
[3]BISAM Central Asia

ABSTRACT

Traditionally Kazakhstan is a country of a transhumant farming system. After the independence from the Soviet Union, state farms were privatized. A large number of newly developed corporate farms were bankrupted through the privatization process, and household farms, a traditional livestock production system using common lands, emerged as private farms. On the other hand, since the prices of general commodities have been raised through the rapid economic growth supported by the oil boom, exporting livestock products becomes more difficult. In addition, it is likely that off-farm work opportunities provided by oil industries have made the traditional farming systems less important in rural economy.

The objective of the study is to quantify the role of the livestock sector in rural society and to elaborate public support strategies in Mangistau Oblast in Kazakhstan. The history of farm restructuring of Kazakhstan and the current trends of livestock sector in Mangistau Oblast were reviewed. Then, the current conditions of restructured farms were examined by household interviews (57 household and 43 peasant farms) and interviews with the leaders of village and corporate farms with regards to land use, production, revenue, marketing, and farmers' attitude.

The results showed that despite unfavorable climate conditions for agricultural production and long distance to market, the rural population in Mangistau Oblast is

* Corresponding author: Wataru Yamamoto, Address: 49 Edgecombe Crescent, Dartmouth NS Canada B2V2A7. Tel/Fax: 1-902-433-1782 Email: wyamamoto99@hotmail.com

strongly attached to their lands and traditional livestock farming. Nearly half of the population belongs to household farms for which 30% of income is supported by livestock production.

Livestock production is largely concentrated in household farms (cattle 90%, horse 78%, camel 76%, and milk 80%). Peasant farms, registered farms with lands of long-term lease, have emerged after struggling for farm restructuring for a decade but the revenue of average peasant farms seems to be narrowly viable. Some corporate farms reorganized voluntarily by farmers were revitalized by new financial support by the state and oil companies, but their future performance is unknown.

The study concluded that organizing livestock farms is important for the convenience of public support to enhance competitiveness in market economy and resource management, particularly at the periphery of settlements. Reorganizing household farms at the village level, formulation of service-oriented associations by peasant farms, development of social entrepreneurship corporation jointly with oil companies, and value addition of livestock products at the village level are suggested. Feasibility studies on such institutions, local-based wool and leather production, and new initiatives to graze rangelands are recommended.

Key words: livestock farm, Kazakhstan, farm organizing, rangeland, farm restructuring.

1. Introduction

Traditionally Kazakhstan is a country of a transhumant farming system. Using nearly 186 million of rangelands, this low Input production system was feasible and supportive to the rural populations (Schillborn van Veen, et al., 2007). During the Soviet Union era, the production systems were centrally controlled and products were systematically sold to prepared market. After the independence from the Soviet Union in 1993, centralized system collapsed and the privatization was introduced. However, the privatization of agriculture through farm restructuring and land reform followed the way of the Cross. Sixty percent of livestock was disposed and a large number of newly organized corporate farms were bankrupted by 1999 (Eserkenov and Beisembaev, 2001; Dudwick, et al., 2007). However, animal inventory of individual farms (household and peasant farms) has actually increased dramatically towards the end of the century (Dudwick, et al., 2007) suggesting that animals were shifted from corporate farms to individual farms. According to the World Bank report, individual farms account for 87% (51% by household and 26% by peasant farms) of livestock production of the country in 2002 (World Bank, 2004) though land reform and farm restructuring in Kazakhstan were supportive of large farms. The private individual farms are considered to be self-motivated and more productive using local resources efficiently, however, they are too small to be competitive in market economy.

In addition, the rapid economic growth supported by oil boom has raised the prices of general commodities; it is hard to compete against foreign producers which produce with lower cost under better climate conditions. Under the powerful oil boom of Kazakhstan, the role of traditional livestock farming may become unclear particularly in Mangistau Oblast where large investment for oil mining, tourism and new town development are jostling (JICA, 2008).

The objective of the study is to quantify the role of the livestock sector in rural society and to elaborate public support strategies in Mangistau Oblast, Kazakhstan. The history of farm restructuring of Kazakhstan and the current trends of livestock sector in Mangistau Oblast were reviewed. Then the current conditions of restructured farms were examined with regards to land use, production, revenue, marketing, and farmer's attitude. The emphasis was placed to compare production and land use of different farm types particularly between household and peasant farms.

2. Study Area

The study area is located in the Mangistau Oblast of Kazakhstan Republic: Latitude 42^0-46^{0}' North and Longitude 50^0 -56^0 East. According to Koppen climate classification, Mangistau Oblast belongs to Desert climate (BW). The climate of Mangistau oblast has continental features: hot and dry summer and cold winter with short transitive seasons, small amount of rains with low humidity, and strong wind and high solar radiation.

The annual average temperatures range from 10^0C to 13.3 ^{0}C. In Aktau (the capital of Mangistau Oblast), the average temperature ranges –0.3^0C in January and 26.8^0C in August. The annual precipitations range between 150mm and 190mm, generally lower in the south, higher in the middle and northeastern parts.

According to statistics data, Kazakhstan has maintained steady growth of GDP around 10% per annum since 2000 (World Bank, 2006). The economic growth is largely supported by export which was annually increased by 27% between 2000 and 2005 and by 37% in 2006 (State of the Republic of Kazakhstan, 2006). Mineral fuel export mainly oil and gas is a driving force of economic growth, accounting for 69% of total export in 2006 (JETRO, 2007). Mangistau oblast is a large contributor for mineral sector in Kazakhstan, accounting for 24% of GDP of mining sector of the country (Statistic Department Mangistau Oblast, 2006). In Mangistau Oblast oil and gas mining accounts for 65% of GRDP. Oil and gas industries provides 33% of employment in 2005, 91% of which are found in urban area (Statistic Department Mangistau Oblast, 2006).

In 2007 the population of Mangistau Oblast is approximately 397,000 people. The population is concentrated in urban areas (Aktau and Zhanaozen) where 73% of total population resides. The population density is 2.3 person/km^2 oblast wide and 242 persons/km^2 at Aktau city.

Mangistau Oblast is composed of two towns (Aktau and Zhanaozen) and four rayons (Beineu, Karakiya, Mangistau, and Tupkaragan). The number of rural settlements in Mangistau Oblast is 52. Regional development program of rural territories in Mangistau oblast classifies rural settlements into three categories based on the level of development priorities. In Mangistau Oblast most of rural settlements belong to medium priority for development, except for Kurik (high priority where a large port development for oil export is planned) and 8 settlements under low priority.

Road network in the rural areas of Mangistau Oblast is poorly developed. The asphalt-paved roads are built only to suburban villages of two large towns, Aktau and Zhanaozen. The majority of rural villages in Mangistau oblast face water deficit and poor quality water

(BISAM, 2007). The majority of rural villages have power supply and 35% has centralized gas supply.

3. Brief History of Land Reform and Agricultural Production in Kazakhstan

The privatization of agriculture in Kazakhstan, as occurred in other CIS countries, was complicated. Agrarian reforms turned out to be extremely painful for overwhelming majority of farmers; the quality of life in rural areas exacerbated drastically through the considerable decline of agricultural production and land deterioration. The situation in rural areas started to be improved at the beginning of the new century when the economic growth began, however, the positive changes in agricultural sector are far less noticeable than in other industries. Social polarization between city and village in Kazakhstan became clear with large economic disparity. A rural area remains as area of the highest poverty rate and lack of development.

In recent years residents of northern oblasts have benefited from grain exports growth with high grain prices. However, residents of livestock regions, including those of Mangistau Oblast, have never gained such benefit since adopted market economy.

The history of agricultural reform of Kazakhstan including those under the Soviet Union is normally divided into the following five stages. After the collapse of the Soviet Union from the second to the fifth stages the reform was aimed at privatization.

1. 1988-1991 By the Law on cooperation in 1988 under the Soviet Union;
2. 1992-1993 Transformation of state property into collective form;
3. 1994-1998 The creation of various forms of corporate and individual management under the mechanism of conditional land and property shares;
4. 1999-2002 Restructure of agrarian sector through mass bankruptcy of corporate farms; and,
5. After 2003 Introduction of private property on agricultural lands.

Agricultural Reform under the Soviet Union (1988-1991)

On the basis of the Law on cooperation in 1988 and some other legislative instruments the first independent peasant farms were established in Kazakhstan. Such farms were established by the most enterprising farmers who still remain as the most successful farms. However, during the Soviet Union era only 1.5% of rural residents have established individual farms disaffiliated from Kolkhozes and Sovhoz.

By the collapse of Soviet Union the agricultural production in Kazakhstan was nearly fully implemented by Sovhoz (state farms) and kolkhozes (officially considered as collective farms, but in fact they were fully controlled by the state). In Kazakhstan more than 90% of agricultural enterprises were Sovhoz, which is much higher than Russia or Ukraine where Sovhoz accounted for less than 40%.

Transformation of State Property into Collective Form (1992-1993)

The actual reforms of agriculture in Kazakhstan started in 1992. The privatization of agriculture was introduced in Kazakhstan in 1992- 93 through the laws "On denationalization and privatization" in 1991 and "On peculiarities of privatization of the property of the state agricultural enterprises" in 1992. The essence of reform was characterized by a peculiar pattern of privatization reflecting Soviet mentality that still dominated in Kazakhstan society. State enterprises were transformed into collective enterprises. As a rule, the former heads of the state enterprises were retained at the positions. Either the forms of partnerships or production cooperatives were established. Peasant farms were allowed to be formed; however, on average less than two peasant farms per each corporate farm were formed. The lands, just as in the Soviet era, were retained as the state property and transferred to agricultural producers for perpetual usage. Thus, at the first stage the denationalization of agricultural enterprises did not develop strong private sector towards the introduction of market economy; agricultural enterprises simply had lost state support.

The Creation of Various Forms of Corporate and Individual Management (1994-1998)

At the third stage (1994-1998) the government tried to rectify the mistake occurred through the transmormation to collective farms by creating a new privatization model. Its key distinction was in personification of a proprietor/owner. The mechanism of conditional land and property shares created various forms of corporate and individual management. However, multiple misuse in the course of privatization, concentration of property in hands of former heads of enterprises during the Soviet era, lack of effective state agrarian policy, impudent plundering of property made majority of established farms nonviable and led to total decline of agrarian production.

In 1994 new transition to privatization was determined by the presidential decree "On additional measures on privatization of property of state agricultural enterprises". By this decree, every worker of a privatizing enterprise received a property share and a conditional land share that determined the right on a certain part of assets and land. There were four ways of managing the shares:

1. to receive real assets (livestock, facilities, etc.) in full ownership or a plot for usage under condition of establishing a peasant farm on the basis of this land and its assets;
2. to contribute the share to an establishing joint stock company, limited liability partnership, collective enterprise or cooperative;
3. to sell or rent it out to other land share owners;
4. to exchange it on land shares of other owners.

In fact only 14% of farmers that were involved in privatization used their shares to establish a farm, while 85% used their shares to join joint stock companies, partnerships, or cooperatives and1% sold their shares (Eserkenov, 1999).

The poor performance of establishing peasant farms was explained mainly by the pressure from the heads of former state farms (reorganizing corporate farms). Farmers were convinced that they wouldn't be able to manage their own farms. Obsolete and almost idle agricultural machinery and non-health livestock were given out to the new peasant farms as a property. It was reported that in some cases of splitting property, taking a half of a tractor wheel or a quarter of a cow was proposed. Those who wanted to get a land to establish a peasant farm received the worst quality lands (BISAM, 1997).

The government actually supported the concentration of property in hands of former heads of Sovhoz since they believed that agriculture would be better managed by experienced managers. In March 1994 a special presidential decree "On transfer of a part of Sovhoz property to ownerships of heads" was issued.

At this stage of the reform the number of peasant farms were considerably increased. By the early 1998 there were about 60,000 peasant farms in Kazakhstan, more than 97% of which were established under the new privatization model after 1994[1].

The third stage of reforms (the second stage under Kazakhstan government) resulted in complete decay of agriculture. Majority of corporate farms were unprofitable by 1998 (Dudwick, et al., 2007). Considerable part of them was insolvent. Between 1993 and 1999 the area under farming by corporate farms shrank by two-thirds, most of which were grazing lands (Dudwick, et al., 2007).

Restructuring Agrarian Sector through Mass Bankruptcy of Agricultural Enterprises (1999-2002)

In 1998-1999 the fourth stage of reforms began, the essence of which was rehabilitation and bankruptcy of enterprises. The government issued "Regulations on peculiarities of an agricultural enterprise bankruptcy procedure". Based on the financial status, all agricultural enterprises were largely divided into three groups: 1) profitable enterprises (Approximately 19% of all the agricultural units); 2) unprofitable enterprises having past-due debts but retaining some liquid property (66%); 3) farms whose bill payable was higher than their assets (more than 15%).

The second group was subjected to rehabilitation procedures, while the third group went to bankruptcy. In fact, by the early 1999 approximately 40% of all the agricultural units were on the list of bankrupts (Eserkenov and Beisembaev, 2001).

Between 1999-2002 agricultural sector in Kazakhstan was seriously restructured. Considerable numbers of farms have lost any rights to property on the former state enterprises, some part managed to take back or take away from agricultural units their land shares and started to rent it out. Thus, land sublease had started.

Introduction of Private Property on Agricultural Lands (after 2003)

The fifth stage of agricultural reforms began in 2003 with adoption of the Land Code that legitimated private property on land of agricultural purpose. This stage has been continuing until present time. The Land Code does not allow irrevocable land transfer to agricultural

[1] Data of the State Agency on Statistics of the Republic of Kazakhstan

producers. Farmer can either buy out the lands as private property, or continue to rent it either for 49 or 99 years. So far an overwhelming majority of farmers have chosen to rent the lands.

According to the Article 170 of the Land Code, the institution of sublease of land shares was eliminated. The holders of rights to conditional land parcel who previously transferred their right for rent to the third persons, were obliged to purchase the land, rent it for 49 or 99 years or transfer it to a shareholder of a joint stock company, limited liability partnership or production cooperative.

About 40% of conditional land shareholders transferred them as contribution to corporate farms. More than 30 percent shareholders transferred or sold their shares to the heads of farms (Eserkenov and Sagin, 2006). By present day the trends of land development in Kazakhstan caused by introduction of private property are not quite clear. The trend to concentration of land property prevails; however, in various rayons it takes place to a variable degree. It is commonly found that property rights were transferred to heads of farms and that the other members became hired workers.

By today the following institutional structure of agricultural productions has been formed:

- Household farms (production on the basis of smallholding having some livestock without their own land)
- Peasant farms (registered individual enterprise with their own lands without the status of legal person, majority of peasant farms consist of member of one or more families).
- Corporate farms (with the status of legal persons provided by legislation of Kazakhstan for any spheres of entrepreneurship – joint stock company, limited liability partnership, production cooperative; the most rapidly developing form is limited liability partnership).

4. Review of the Livestock Sector in Mangistau Oblast

Land Resources

Mangistau Oblast covers approximately 16 million hectare. Beineu, Karakya and Mangistau rayons account for 92% of total area (Table 1). Approximately one half of the area is used for agriculture (Livestock production), 83% of which is pasture. Sixty-seventy percent of land is used for livestock production in Karakia and Mangistau and Tupkaragan, while only 6.8% of land is used in Beineu. As a whole 40% of land is under land reserve, mostly unoccupied rangelands traditionally used for summer pasture. Approximately 20-30% of area is under land reserve in Karakia, Mangistau, and Tupkaragan rayons because of Ustyurt plateau which covered the eastern part of the oblast. Large land reserve (82 % of total area) is found in Beineu rayon due to the appearance of salt in the soil.

Table 1. Land resources in Mangistau Oblast

	Total area		Settlement		Agriculture land (ha)						Land reserve(ha)					Others(ha)	
Rayon/city	(ha)	%	(ha)	%	Total area	%	Arable	Hay	Pasture	Others	Total area	%	Arable	Hay	Pasture	Area	%
Beineu	4,051,933	24.5	230,940	5.7	276,039	6.8	0	100	268,647	7,292	3,308,671	81.7	4	186	2,344,661	236,283	5.8
Karakiya	6,329,262	38.2	240,741	3.8	3,819,092	60.3	36	0	2,906,921	912,135	1,995,467	31.5	0	0	1,608,979	273,962	4.3
Mangistau	4,789,122	28.9	190,412	4.0	3,447,163	72.0	104	0	3,170,411	276,649	1,096,311	22.9	0	0	434,450	55,236	1.2
Tupkaragan	1,259,596	7.6	33,423	2.7	891,997	70.8	158	0	675,627	216,212	252,426	20.0	0	0	203,289	81,750	6.5
Aktau	80,629	0.5	21,941	27.2	3,438	4.3	113	0	2,454	871	4,408	5.5	0	0	0	50,842	63.1
Zhanaozen	51,549	0.3	9,011	17.5	331	0.6	100	0	13	218	0	0.0	0	0	0	42,207	81.9
Total	16,562,091	100.0	726,468	4.4	8,438,060	50.9	510	100	7,024,072	1,413,377	6,657,283	40.2	4	186	4,591,379	740,280	4.5

Source: Village monitoring: Mangistau Oblast. 2006.

Farm Structure

There are approximately 33,000 farms in Mangistau oblast 48% of which is household farms (grazers) (Table 2). Household grazers do not have their lands. Most of the area (88%) is owned by either limited liability partnership (65%) or cooperatives (24%), while areas of peasant farms account for 11%. The average size of peasant farms is 962ha. Under the Soviet Union, there were 21 Sovhoz but no Kolkhoz in Mangistau oblast.

Table 2. Area and average size by farm types in Mangistau, 2007

Farm type	Number	Area (ha)	% of area	Average size (ha)
Public Enterprises	7	283	0.00	40
Limited Liability Partnership	73	5,330,793	64.6	73,025
Joint Stock Company	3	10,233	0.1	3,411
Production Cooperative	38	1,961,279	23.8	51,613
Peasant farm	979	942,001	11.4	962
Household grazers	30,615	2,586	0.03	0.08
Urban	12,764	929	0.01	0.07
Rural	17,851	1,657	0.02	0.09
Others	1299	193	0.00	17.76
Total	33,014	8,249,954	100.0	129.66

Source: Department of Statistics, Mangistau Oblast.

Animal Inventory

Animal inventories were compared between 1991, 2002 and 2006 (Table 3). The large declines between 1991 and 2002 as well as the recovery between 2002 and 2006 were observed for all type of animals. The number of camels in 2006 was larger than those of 1991. Animal inventory in 2006 for sheep, horses, and camels were 71%, 76% and 110% of 1991, respectively. It should be noted that the numbers of sheep and horses in Aktau and Zhanaozen in 2006 are much larger than those in 1991, suggesting that animals are gradually concentrating near the towns.

Livestock Production

As a whole, in 2006 meat production by peasant farms increased, while those by corporate farms decreased (Table 4). Regarding the production by animal type, cattle, pig, and horse production have increased, while sheep and camel have decreased in 2006. Although the amount is rather small, the increased cattle production by household farms and pig production by peasant farms were significant (102% and 432% increase, respectively). Milk production is has slightly increased in 2006 largely by the contribution of peasant farms.

Milk production is largely dominated by household farms. Milk is produced almost exclusively by household and peasant farms 80% and 18%, respectively. The most common dairy products in Mangistau is Shubat, sour milk produced from camel milk. As of the beginning of 2007, there are 20,800 female camels are kept in Mangistau oblast, 35-37% of which are milked. Camel produces milk from May to November. In the production at local farms, the quality of Shubat deteriorates in summer. Cow milk is exclusively produced at household farms with low productivity due to the lack of fodder (Agriculture Department Mangistau Oblast, 2007).

Table 3. Animal inventory in Mangistau Oblast

Animal	Year Rayon/city	1991	2002	2006	% of 2006 for 1991
Sheep	Aktau	3,377	4,942	8,925	264.3
	Zhanaozen	4,990	3,353	6,593	132.1
	Beineu	169,992	79,030	118,486	69.7
	Karakiya	110,270	61,225	96,532	87.5
	Mangistau	286,103	180,273	187,618	65.6
	Tupkaragan	75,330	37,404	42,620	56.6
	Total	650,362	366,227	460,774	70.8
Horse	Aktau	161	330	593	368.3
	Zhanaozen	296	459	369	124.7
	Beineu	6,773	3,802	5,600	82.7
	Karakiya	9,857	4,391	4,801	48.7
	Mangistau	16,716	12,201	15,444	92.4
	Tupkaragan	7,164	4,435	4,353	60.8
	Total	40,767	25,618	31,160	76.4
Camel	Aktau	25,254	707	832	3.3
	Zhanaozen	820	907	1,198	146.1
	Beineu	7,710	7,335	10,128	131.4
	Karakiya	7,412	3,748	5,080	68.5
	Mangistau	10,234	11,140	14,708	143.7
	Tupkaragan	7,439	5,199	5,397	72.6
	Total	33,869	29,036	37,343	110.3

Source: Department of Statistics, Mangistau Oblast.

The meat production in Mangistau in 2006 is sheep (54%), camel (21%), horse (19%), cattle (6%), and pig (0.4%) (Table 4). As occurred in milk production, meat production is largely dominated by household farms, accounting for 90%, 56%, 78%, and 76%, for cattle, sheep, horse, and camel, respectively (Table 5). Although the amount was still small, peasant farms was dominant in pig production in 2006 (90%), increased dramatically from the previous year (Table 5). Corporate farms focus more on sheep production (23%, Table 5). It was estimated that in 2006 approximately 9,600 horses are annually slaughtered accounting for 31% of stock with average of 180kg, while 9500 camels slaughtered for 25% of stock with average weight of 210k.g. As a whole, 92% of meat are produced at either household or peasant farms.

Wool are also largely produced at household farms (60%) (Table 4). The volume of wool processed in Mangistau oblast accounts for only 10% of production potential (Agriculture Department Mangistau Oblast, 2007).

Table 4. Livestock production in Mangistau Oblast, 2006 (Unit: ton)

Farm type	Total		Corporate farm		Peasant farm		Household farm	
Products	2006	Increase (%)	2006	Increase (%)	2006	Increase (%)	2006	Increase (%)
Live weight (Total)	9483.6	98.7	1528.3	90.9	1650.0	100.1	6305.3	100.5
cattle	594.1	150.9	1.5		57.1	44.2	535.5	202.4
sheep/goats	5100.6	93.3	1191.1	93.9	1030.3	106.3	2879.2	89.2
pigs	34.5	271.7	0.6		31.4	532.2	2.5	36.8
horses	1756.8	115.1	158.6	84.3	231.0	100.8	1367.2	123.0
camels	1988.7	90.5	176.5	78.3	300.2	95.1	1512.0	91.3
Milk - total ton	6055.4	102.3	127.6	88.9	1068.4	121.8	4859.4	99.2
Wool - total ton	773.0	103.7	158.8	85.6	150.6	111.0	463.6	109.3
Hide. pc	195,788	288	60,042	253	34,331	326	101,415	310

Source: Department of Statistics, Mangistau Oblast. Note: Increase is shown by percentage of 2006 in comparison with 2005.

Table 5. Percentage of production by farm type

	Household farm	Peasant farm	Corporate farm
Cattle	90	9.6	0.3
Sheep/goats	56	20.2	23.4
Horses	78	13.1	9.0
Camel	76	15.1	8.9
Pig	7	91.0	1.7
Milk	80	17.6	2.1
Wool	60	19.5	20.5

Table 6. Land use by farm type

	Peasant farm		Household grazer	
Land type	Area (ha)	Number	Area (ha)	Number
Pasture	1276	44	0	56
Crop	12	9	2.8	5
Residential	1799 m^2	44	1278 m^2	56

Rural Finance for Agricultural Development

In Kazakhstan micro credit for rural populations is provided by Agriculture Financial Support Fund (Joint stock company owned by KAZAGRO, national holding company to support agricultural development). In 2006 Agriculture Financial Support Fund provided credit of 71.1 million Tenge (3 times more than micro-credit given in 2005) for 559 households. The credit is provided without any collateral, but the guarantee by Rayon Akim (Rayon leader) is required. The amount of credit per person in 2007 is up to 800,000 Tenge (US$6,600) for juridical persons and 260,000 Tenge (US$2,160) for physical persons. The limit of the credit is planed to be raised up to 2 million Tenge (US$17,000) and 500,000 Tenge (US$4,170), respectively. The borrowing conditions in 9.5% of annual interest with 3 years of payback period. According to an officer of Agriculture Financial Support Fund, most credit is paid back without delay (96%) as of the end of 2007.

Agriculture Financial Support Fund financed three other micro-credit organizations (Uzen, Beineu and Munalinsky) in 2007 in order to disseminate micro-credit to rural areas.

5. Farm Survey

Methodology

Farm surveys were conducted in five villages (Shaiyr, Kyzylozen, Tourysh, Senek, and Zhynguldy), one village for each rayon except for Mangistau rayon (Shaiyr and Zhynguldy). The typical villages having sufficient number of populations to provide representative samples for the surveyed rayon were selected. For the survey structured interviews were applied. Twenty families were interviewed for each village (100 interviews in total, 57 with household, and 43 peasant farms) in October, 2007.

Household and peasant farms in the each village were selected by route sampling. In each village a head of a household/farm (if possible) or another person capable of giving the most complete information was interviewed. The samples were selected only from household/peasant farms that produce and sell livestock products.

Farm History and Land Use

Most of the surveyed farms in Mangistau oblast had long history of settlement; 80% of farmers had lived in the villages since the settlement was created. Almost all the surveyed farms (95%) lived in the same lands as present residence during the Soviet Union era. No farm was registered as a peasant farm before 1991. Two thirds of farms were registered after 2000. Overwhelming majority (95%) of peasant farms received the land on the basis of land share.

On average peasant farms had 1276 hectare of pasturelands (Table 6), which was larger than Oblast average. Nine farms had crop fields of 12 hectare on average. Household farms did not have any pasturelands, but some farms did have small crop fields. All the farms owned residential area, and rented pasture and crop fields from the state.

Farm Income and Economic Activity

The annual revenue of households included income from agricultural products, wages of family members, pensions and work in other organizations. The livestock production was the major economic activity for majority of peasant farms (70%), and employment income for household farms (54%) (Table 7). The livestock products were main income source for approximately one third of household farms. The income from livestock production accounted for 100% at 8 farms.

On average farmers received annual revenue of 1.24 million Tenge (US$10,300, 1US$= 120 Tenge in 2008) (1.52 million Tenge for peasant farms and 1.02 million Tenge for household farms), the highest was reported at Zhyngyldy village in Mangistau rayon and the lowest from Kyzylozen village at Tupkaragan rayon. Livestock production accounted approximately a half on average in peasant farms and 30% by household farms (the highest annual income 7 million Tenge and the lowest 350,000 Tenge were reported.).

Table 7. Major economic activities by farm type

	Peasant farm		Household farm		Total	
Major economic activity	Number	%	Number	%	Number	%
Livestock production	31	70	19	34	50	50
Hired at extracting company	3	7	7	13	10	10
Crop cultivation	4	9	0	0	4	4
Hired at other organization	6	14	30	54	36	36
Total	44	100	56	100	100	100

Table 8. Average farm income at surveyed village

	Peasant farm		Household farm		Total	
Village	Revenue	%	Revenue	%	Revenue	%
Shaiyr	1,725,714	64	1,1271,54	20	1,336,650	36
Kyzylozen	1,232,683	50	871,571	28	979,905	34
Tourysh	969,986	85	1,067,346	40	1,033,270	56
Senek	1,421,500	47	800,545	36	1,079,975	41
Zhyngyldy	1,870,180	35	1,562,940	6	1,793,370	28
Total	1,525,277	52	1,024,129	29	1,244,634	39

Note: N=100. % is the percent of income from livestock production for total income.

Herd Structure and Livestock Production

The survey results showed that sheep were owned by all the peasant and household grazers, camels by most of farms (82% and 88% respectively) but horses were owned at larger extent by peasant farms than households (84% and 41%, respectively). Cattle were owned by fewer farms (a half of peasant farms and 36% of household farms).

Sheep/goat were produced for meat and wool at all the farms, camel for milk and less extent for meat, cow for milk with limited amount (Table 10). Almost all the farms (95%) sold live animals (Table 10).

Average liveweight of animals reported by the farms were 24kg for sheep, 207kg for camel, 217 for cow, 163kg for horse. Average price of the animals were 10,184 Tenge for sheep, 123,429 Tenge for camel, 86,667 Tenge for cow and 96,667 Tenge for horse. Meat price was approximately 550 Tenge per kg for all the animals.

Table 9. Herd structure by farm type

	Peasant farm			Household farm		
Animals	Number	Farms	%	Number	Farms	%
Sheep/goat	238±34	44	100	33±3.0	56	100
Horse	11.3±2.1	37	84	2.4±0.5	23	41
Camel	8.9±1.8	36	82	4.2±0.6	49	88
Cattle	3.3±1.4	22	50	0.8±0.2	20	36

Note: Mean± Standard Error: N=44 for peasant farms, N=56 for Household farms.

Table 10. Livestock production of the last year by farm type

		unit	Peasant farm			Household farm		
Animal	Product		Production	# of farms	%[1]	Production	# of farms	%[1]
Sheep	Leather	kg	18±2.6	44	100	11±1.2	56	100
	Wool	kg	457±81.9	44	100	59±6.6	56	100
	Milk	Liter	41±21.3	2	5	0.1	1	2
	Meat	kg	1465±826	44	100	251±84	54	96
Horse	Leather	kg	0.6±0.1	16	36	0.4±0.2	14	25
	Wool	kg	0.3	1	2	0	0	0
	Milk	Liter	0	0	0	0	0	0
	Meat	kg	103±26.7	18	41	35±6.0	15	27
Camel	Leather	kg	0.8±0.1	22	50	0.5±0.2	19	34
	Wool	kg	47±10	37	84	14±3.1	45	80
	Milk	Liter	3589±1444	35	80	859±230	44	79
	Meat	kg	192±48	23	52	115±97	15	27
Cattle	Leather	kg	0.1±0.1	4	9	0.1	1	2
	Wool	kg	0		0	0	1	2
	Milk	Liter	23±42	3	7	111±90	10	18
	Meat	kg	14±7.8	6	14	3.6±4.1	3	5

Note: Mean± Standard Error: N=44 for peasant farms, N=56 for Household farms.

[1] Percent of farms to produce the products for total number of farms.

Approximately, 80% of the surveyed households were engaged in camel milk production, while some farms (7 and 18% of peasant and household farm, respectively) produced cattle milk. Majority peasant and household farms (80%) produced camel milk and 66% of farms sell milk with 130 Tenge per litre. Most of milk produced in the region is consumed at home.

Typical farm income from livestock products based on mean production of meat and milk and market price were estimated as 1,450,000 Tenge (US$12,000) for peasant farm and 350,000 Tenge (US$2,900) for household grazers. Majority of farms (82%) were interested in increasing livestock production, but most of the farms (77%) thought that the production level was the same as the previous year.

Marketing and Attitude for Reorganizing Farms

Majority of the farms sold their products through bulk buyers or by carrying the products to town or rayon centers. The role of bulk buyers in the marketing system was larger for peasant farms (43%) than household farms (20%) (Table 11). Majority of peasant and household farms (75% and 57%, respectively) were not interested in reorganizing farms, while 39% of household farms were interested in reorganizing farms (Table 12).

Table 11. Marketing methods by farm type

	Peasant farm		Household farm	
	#	%	#	%
Bulk buyer	19	43.2	11	19.6
Local middlemen	4	9.1	1	1.8
Local market by themselves	3	6.8	1	1.8
Carry to town/rayon center	25	56.8	44	78.6

Table 12. Interest of farmers in reorganizing

	Peasant farm	%	Household farms	%
Quite interested	2	4.5	1	1.8
Tend to say interested	3	6.8	22	39.3
Neither nor	2	4.5	0	0.0
Tend to say not interested	33	75.0	32	57.1
Not interested at all	4	9.1	1	1.8
Total	44	100.0	56	100.0

Animal Health, Pasture Conditions, Growing Other Plants

The survey results showed that animal health conditions in Mangistau oblast were rather well; 82% of the surveyed farms noted that for the last 5 years the livestock didn't suffer from any serious diseases, but 80% of farm reported a problem of insects. However, only 61% of farms vaccinateed animals.

Majority of farms (86%) use permanent pasture, while 14% used seasonal pasture: summer and winter pastures. Majority of farms (96%) reported the need of pasture, but only 9% cultivated them. Overwhelming majority of farms (91%) purchased winters fodder. Agricultural crops were cultivated by 14% of the surveyed farms. Eight farms sold crops at market outside of villages with price of 46 Tenge per kg. Majority of respondents did not cultivate plants, except for Tourysh village where a half of farms grew plants. On average one third of farms (36%) planted trees.

6. Case Studies: Village and Corporate Farms

The case studies were carried out for one village (Jynghyldy) and two Limited Liability Partnerships (LLP) by leader interviews. The village was selected from livestock active region, while LLPs were selected from active ones.

6.1. Jynghyldy Village

Jynghyldy village in Mangistau Rayon was established in 1992 separated from Kolkhoz (359,395 ha). The village has territory of 27,136 ha with 2281 people in 427 households (Table 13). There were three types of farms in the village: 363 household farms grazing mainly village pasture (communal grazing lands within the village territory); 64 peasant farms (which were separated from cooperative) and one large production cooperative (with forty workers) which is the remnant of the Kolkhoz.

On average one household in the village had approximately 53sheep/goat and a few horses and camels (Table 14).

Table 13. Land use of Jynghyldy Village

Land Type	Area (ha)
Settlement	319
Salty field	1,223
Pasture	25,594
Total	27,136

Table 14. Farm number and area by farm type

Item/Farm type	Household farm (Total)	Household farm (Average)	Peasant farm (Total)	Peasant farm (Average)	Production Cooperative
Number of farm	363		56		1
Land (ha)	25,494	70	74,602	1332	227,000
Cattle	180	0.5	57	1.0	-
Sheep	13,636	37.6	6395	114.2	4004
Goat	5,050	13.9	919	16.4	-
Horse	675	1.9	898	16.0	25
Camel	1,204	3.3	479	8.6	19
Poultry	291	0.8	19	0.3	
Livestock unit (LU)	5463.9	15.1	2770.6	49.5	654.1
Stocking Rate (SR)	0.21	0.21	0.04	0.04	0.0029

Source: Akimat Jynghyldy village.

Note: LU (200kg): based on 200kg cattle/horse, 30kg sheep/goat, and 300kg camel.

Peasant farms had 49 year rent agreement for their farm lands based on the land code 2003. Stocking rate calculated based on village area suggests that the stocking rate of household farms was 6 times higher than peasant farms and 74 times higher than cooperative. It may suggest that household farms are far more productive than peasant farms, while cooperative has extremely low productivity. The productivity can be potentially increased at peasant farms, but overgrazing may occur near village settlement. It was likely that the cooperative had lost incentive for better management.

6.2. Limited Liability Partnership

LLP Taushik Aubil Sharuashiligi

Limited Liability Partnership (LLP) Taushik Aubil Sharuashiligi was established in the area of former Solvoz in Taushik village. The land area of 420,000 ha is leased for 49 years in 2001. The Solvoz was separated from village settlement area, peasant farmlands and the company's leased land. LLP Taushik Aubil Sharuashiligi has approximately 200 employees organized by 55 farm units. It has a capital of 18 million Tenge from 53 shareholders; most of them work for the company.

It has approximately 3000 camels, 500 horses and 500 sheep. Camel milk is sold a half to Kalamkas oil field and the rest for Aktau. One hundred twenty tons of Shubat is produced annually. It produces 12 tons of camel fur and sells to a local business man in June.

LLP Taushik Aubil Sharuashiligi is building a factory to make canned camel meat with credit of KazAgrofinance (10 million Tenge with 9% interest). It is difficult to increase the number of camels due to the high price of supplemental feeds in the winter.

LLP Senek

LLP Senek was established in 1997 maintaining an exact form of Sovhoz. LLP Senek has approximately 3 million ha and 40,000 sheep, and 700 horses with 400 families. LLP Senek annually produces 120-150 tons of meat and 30-35 tons sheep wool for insulation materials for oil company UzenMunaiGas. UzenMunaiGas supported LLP Senek investing approximately 1 million Tenge to purchase equipment for skin processing. The skins are currently sold to Russian. LLP Senek expects to produce 15000 processed hides with US$15. LLP Senek uses summer pasture on Ustyurt Plateau for all animals.

7. DISCUSSION

Viability of Livestock Sector in Rural Economy

Despite the economic growth by oil industry, livestock production remains as the major economic activities for the rural livelihood in Mangistau Oblast. There are approximately 30,000 household farms in Mangistau oblast. Based on the average family size (6 member per family), it suggests that nearly a half of population (almost all the population in rural area) belongs to household farms. The survey results showed that livestock production is main income for approximately 70% of peasant farms and 54% of household farms. The annual income from livestock products (1,450,000 Tenge (US$12,000) for peasant farm and 350,000 Tenge (US$2,900) for household grazers) estimated based on mean size of farms is large enough to survive for a low input production system in rural society considering the average monthly wage of Mangistau Oblast (64,000 Tenge (US$530)) though production cost was not subtracted in this study. Rural populations in Mangistau oblast have long history of settlement, and the majority of farmers (82%) demonstrated their intension in increasing livestock production.

Farm Restructuring

Land reform of Kazakhstan was described as the combination of having share of privatization, reducing the size of farms, and restructuring cooperate farms, and forming peasant farms (Dudwick, et al., 2007). Both peasant and corporate farms rent state lands, and corporate farms have the status of legal person and capital developed by members. Household farms are animal owners who graze in village territories. The case study from Jynghyldy village showed that the stocking rate of household farms was much higher than peasant farms, and that of the remaining collective farms was extremely low.

In order to revitalize livestock production in Mangistau oblast, the farms need to be competitive in market economy. In order to maintain the competitiveness of the rural products, farmers also need to collaborate each other particularly when the distance is a problem. Since the capital market in rural areas is not well established in Kazakhstan, the government's role is to develop financial mechanism, introduce modern technologies, and provide market information to support farms. The lack of government institutions for the support of household farms and a lack of rural finance resource for small farms were suggested (Dudwick, et al., 2007). Organizing producer's units is needed in order to make public supports more effective.

Livestock Production by Farm Type

The production of cattle, pig and horse increased in 2006 (Table 2). The increased productions were largely contributed by household farms, except for pig which is contributed by peasant farms. Camel production were reduced but the actual number was increased in 2006 compared with 2002 (Table 4), but the inventory differed largely by region. The inventory of camels were larger in rural area in general.

Animal inventory in 2006 recovered dramatically compared with 2002 (Table 3). However, compared with 1991, camel inventory is already higher, but sheep and horse are still at the level of 71% and 76%, respectively. The animal inventory of corporate farms reduced dramatically after the collapse of the Soviet Union. It suggested that the animals shifted from corporate farms to household and peasant farms as occurred in other Oblast (World Bank, 2004). Currently household farms accounts for 66% of animal inventory, while peasant and corporate farms account for 17% and 16%, respectively.

Reconsolidation of Household Farms into Village Farms

A large part of livestock production in Mangistau Oblast is produced by household farms (90% of cattle, 78% of horse, 76% of camel and 80% of milk). Currently there are approximately 30,000 household grazers. Livestock production is a part of their income for household grazers, but most of them have other source of income. The support or restructuring of these farms are the key problem of livestock sector in Kazakhstan. It should be noted that household farms existed during the Soviet era as each household has their own animal in state farms. As seen in the case of Jynghyldy village, the circumference of village territories tends to be overgrazed due to high stocking rate of the households, potentially causing pasture degradation and desertification in many villages. The problem of desertification was pointed out as one of major environmental problem at almost all the villages (BISAM, 2007).

In order to properly manage the large number of livestock and effectively introduce a modern production system (e.g. grazing management, optimization of nutrition, market-oriented selection of productive breed, product diversification, value addition and marketing), households should be reconsolidated at village level. Animal husbandary in common territory tends to exploit natural resources. The study results showed that approximately 40% households are interested in reconsolidating their farms. The share of each household for village farms can be based on the numbers of livestock owned. Currently according to the Land code 2003, the lands previously used by local population as pasturelands cannot be given in shared ownership (Schillborn van Veen, et al., 2007). Thus the new land code cannot arrange for land use around the settlements. Therefore, these common lands estimated as 17 million ha nation wide are not managed and unprotected against overgrazing (Schillborn van Veen, et al., 2007).

Formulation of Service Oriented Associations to Support Peasant Farms

Currently there are approximately 1,000 peasant farms in Mangistau Oblast. After the collapse of the Soviet Union, in the process to the transition to market economy through land reform, peasant farms are considered to be one of the most productive units for livestock production in Kazakhstan. However, due to lack of capital and technologies, the peasant farms have problems of inappropriate breeds for market, low stocking rate, and little or no value addition to raw material. In order to be competitive in market economy, peasant farmers need to be strengthened. The study results showed that only 15% of peasant farmers are interested in reorganizing them because of unfavorable experience during the Soviet era. Service oriented associations by peasant farm dealing with product marketing, machinery leasing/purchase, collaborative material purchase and products sales, breed replacement, veterinary service, construction of warehouses, credit delivery, etc. can be organized (Lerman, et al., 2004), but entrepreneurship-oriented approach should be taken since reorganizing farms makes farmers imagine to form bankrupted corporate farms again.

Potential Collaboration with Oil Industry

The surveyed farmers are aware that the situation can be changed through construction of the storage, butchering facilities, refrigerators, and local wool and leather processing workshops. However, they rely on the state support when it comes to such practices. They expressed no initiative in creation of associations for managing product quality improvement or marketing. The supports for such purchase/construction are not planned by the

Governmental Program on Rural Development in Mangistau Oblast. In order to justify the investment on the development of livestock industries, the farm outputs need to be competitive against local manufactures which use raw materials of better quality delivered from other regions of Kazakhstan and from abroad. In order to support these farms in market economy, the program for the development of socio-entrepreneurship corporations collaborated with oil industry may be implemented.

Organizing Rural Micro Credit Organization for Comprehensive Support to Farmers

One way to effectively establish public support for local farmers is to use rural finance mechanism. Micro credit provided by Agriculture Financial Support Fund has increased three times larger in 2006 than 2005. Micro-credit organizations were established in rural areas including Uzen, Beineu, and Munailinsky in 2007. In 2008, micro credit organizations are planned to be established at each rayon in order to provide micro credit from the fund provided by Agriculture Financial Support Fund.

The new micro credit organizations established in each rayon may be developed as multi functional farmers' organization to connect farmer's needs with public support. The new farmer's organization can not only provide micro credit, but also support other things (e.g. buying raw materials, selling products, receiving subsidies, providing technical training). In addition, the farmer's organization can eventually function as rural banks; farmers' additional money can be saved at the farmer's organization as seen in agricultural cooperatives in other countries[1].

Market Development Connected to Town Center

The majority of the surveyed peasant farms sell live animals. The meat is sold by approximately a half of the household/peasant farms, mainly at local markets in small quantities. The camel milk is mostly consumed at home and is sold in small quantities only by 35% of households/peasant farms.

According to the rural survey, most of livestock products are sold at markets in the nearest city or rayon center (56%) or through bulk buyers (31%). It is likely that sheep meat produced in Mangistau Oblast is not much sold in Aktau market due to unfavorable quality than those from Kostanai and Uralsk. Market channel in Aktau city with local farmers needs to be established.

Introduction of Small Scale Value Addition of Livestock Products at Village Level

Due to the economic growth by oil production, the price of raw materials produced from animals has declined in Kazakhstan. As a result, a large percentage of hide and wool of sheep

[1] For example, Kerala State Cooperative and Agricultural Bank in India and Agriculture cooperative in Japan.

are wasted in Mangistau Oblast. The difficulty to sell hide and wool was reported as the most serious problem at surveyed households. Most of hide and wool were simply thrown away due to the lack of market. The difficulty of wool sales are caused by low quality (rough and dirty) as well as lack of processing and marketing network. Value addition for wool and leather (e.g. improvement of camel wool quality, washing of Karakul wool) are expected to be introduced in order to increase livestock sales. Feasibility study on small-scale handicraft manufacturing for wool processing is recommended.

Fodder Availability and Use of Rangelands

In order to increase productivity, winter fodder is a problem. Most of farmers choose to purchase winter fodder, without using summer grazing since centralized livestock moving control collapsed (Alimaev and Gehnke, 2008). Local based hay production should be encouraged; however, it should be noted that still 14% of farmers are using summer pasture. Also LLP Senek is systematically using summer pasture in a large scale. Since supplementary feeding with crops are costly, utilization of summer pasture is still considered to be important in order to preserve pasture for winter. It is likely that some parts of Ustyurt plateau has sufficient water source and animals are being grazed. Further study on use of rangelands under market economy particularly with reorganizing farms is recommended.

CONCLUSION

Despite unfavorable climate conditions for agricultural production and long distance to market, the rural population of Mangistau Oblast is strongly attached to their land and traditional livestock farming. Nearly a half of population belongs to household farms of which 30% of income is supported by livestock production. The majority of farmers are interested in increasing the number of animals and to produce and sell livestock products.

Although the number of livestock has declined through farm restructuring process in the 1990s after the collapse of the Soviet Union, the increase of number of animals took place during recent years. However, reactivated livestock production is simply concentrated on household grazers, traditional livestock production systems by rural residents using common lands. Household grazing is dominant in cattle, horse, camel, and milk production.

Although the number is limited, peasant farms, registered farms with lands of long term rent emerged after struggling for farm restructuring for a decade. The average revenue of peasant farms seems to be narrowly viable. Some corporate farms reorganized voluntarily by farmers under market economy were revitalized by new finance support by the state and oil companies, but their future is unknown at this stage. In order to effectively maintain livestock production reorganizing livestock farms is important for both for public support to make them competitive in market economy and resource management. Formulation of village farms by organizing household grazers, development of service-oriented associations particularly through the dissemination of micro credit, developing small-scale social entrepreneurship corporation with oil companies are proposed. Feasibility studies on such institutions, local-based wool and leather production, and new initiatives to graze rangelands are recommended.

ACKNOWLEDGEMENTS

The study was benefited by the project, "Master plan study on Integrated Regional Development for Mangistau Oblast, the Republic of Kazakhstan" funded by Japan International Cooperation Agency (JICA). The authors wish to thank for providing the opportunity to undertake this study.

REFERENCES

Agriculture Department, Mangistau Oblast. (2007) The program of prime measures for 2007-2008 on realization of the Program for steady development of agriculture Mangistau for 2006-2010. Программа первоочередных мер на 2007-2008 годы по реализации Программы устойчивого развития агропромышленного комплекса Мангистауской области на 2006-2010 годы.

Alimaev, I. and Gehnke, R. Jr. (2008) Ideology, land tenure and livestock mobility in Kazakhstan. In K. A. Galvin et al. (eds.), *Fragmentation in Semi-Arid and Arid Landscape: Consequences for Human and Natural Systems*. Springer.

BISAM (1997) Privatization and Restructuring of Agriculture in Akmola and Former Taldy-Korgan Oblast. World Bank.

BISAM. (2007) Survey on rural development in Mangistau Oblast. The master plan study in integrated regional development project (IRDP) for Mangistau Oblast. *JICA.*.

Dudwick, N., Fock, K., and Sedik, D. (2007) Land Reform and Farm Restructuring in Transition Countries: The experience of Bulgaria, Moldova, Azerbaijan, and Kazakhstan. *World Bank Working Paper*. No. 104. World Bank.

Eserkenov, T. (1999) Privatizatsiia gosudarstvennoi sobstvennoisti v Respublike Kazakhstan v usloviiakh perekhoda krynku (*Privatization of state property in the Republic of Kazakhstan under conditions of transition to market*.) Almaty.

Eserkenov, T. and Beisembaev, T. (2001) Bankrotstvo sel'skoshoziaistevennykh predpriiatii: teoriia i praktika. (*Bankruptcy of agricultural enterprises: theory and practice*.) Almaty.

Eserkenov, T.and Sagin, Zh. (2006) Improvement of land relations in the Republic of Kazakhstan under market conditions. Kokshetau.

JETRO (2007) *Global Trade Atla*s.

JICA (2008) Master plan study on Integrated Regional Development for Mangistau Oblast, the Republic of Kazakhstan.

Lerman, Z, Csaki, C. and Feder G. (2004) *Agriculture in transition: Land policies and evolving farm structure in Post-soviet countries.* Lexington Books.

Schillborn van Veen, T.W., Alimaev, I.and Uakelov, B. (2007) Kazakhstan: Rangeland in transition, the resource, the users and sustainable use. *World Bank Technical Paper*. World Bank.

State of the Republic of Kazakhstan. (2006) *Kazakhstan for years of independence*.

Statistic Department Mangistau Oblast. (2006) *Statistical Yearbook of Mangistau Ob*last, 2006.

World Bank. (2004) Kazakhstan's livestock sector – Supporting its revival. *The Kazakhstan Government and World Bank*.

World Bank. (2006) *World Development Indicators Database*.

In: Agricultural Systems: Economics, Technology and Diversity ISBN 978-1-60692-025-1
Editor: Oliver W. Castalonge

Chapter 6

CAN THE SUSTAINABLE LIVELIHOOD FRAMEWORK BE QUANTIFIED? THE DIVERSE LIVELIHOODS IN THE TAITA HILLS OF KENYA

Eija Soini*
World Agroforestry Centre (ICRAF), Nairobi

ABSTRACT

The Sustainable Livelihoods (SL) framework is widely used for structuring qualitative studies of livelihood. This chapter tests the feasibility and value of doing a quantitative survey in the Taita Hills of Kenya using the SL framework. The framework was used as a structure both in the quantitative data collection and the statistical analysis of interrelationships of livelihood assets and outcomes. According to the survey, some of the biggest changes in the Taita Hills, as commonly perceived by farmers, are the declining soil fertility, decreasing number of livestock, new agricultural technologies, changes in the number of trees and species, and the increasing need of off-farm income. There is great variation in perceived changes, problems, adaptation strategies, coping strategies with unexpected expenditures, experimenting with new activities, and achieving desired livelihood outcomes. To study interrelationships of the different livelihood capital and livelihood outcomes several variations of multivariate regression analysis were used. The livelihood objectives that were considered by the farmers themselves as the most desirable in improving the standard of living were used to select the components of the livelihood outcome. Although the process is dynamic, with each household following a different time trajectory, it is expected that some links between the capital and the outcomes can be detected in cross-sectional data. The linear multivariate regression analysis showed that only financial capital could be directly related to the measured livelihood outcomes. But the variance accounted for is small, 26%. Analysing the components of outcome and capital separately reveal some patterns, and several of these challenge conventional expectations. However, the overall picture is one of much variation and individuality in livelihoods. The SL framework was found useful in the design of the survey. It focuses the survey instrument on the key elements, and it helps structure the presentation and analysis. However, it has some shortcomings when used

* E-mail: eija.soini@iki.fi

with quantitative data. Much remains unexplained by a quantitative analysis using the sustainable livelihood SL framework. This may be due to the multiple livelihood strategies, selection of indicators, and/or the many intangible assets and social processes not described by the framework. However, despite the great variation, financial capital may be the most crucial asset in influencing the realization of livelihood outcomes. Increasing occupational multiplicity is a general trend in rural Africa. The increasing need of cash and the limited income from farm production is forcing farmers to look for off-farm jobs to supplement farm income.

Key words: Taita Hills, livelihoods, sustainable livelihood framework, East Africa, highlands, adaptation to change

INTRODUCTION

The Taita Hills, part of the Eastern Arc mountains, is one of the highland areas in East Africa which has gone through significant change in land use over the past decades. Population has increased considerably through in-migration and natural growth, and spread to the adjacent lowlands. Large sisal estates and National Parks have been established in the lowlands. Government land consolidation policies implemented over several decades have changed the farming system and the livelihood strategies (Nazzaro 1974; Fleuret 1988; Soini 2005). The Taita Hills is also one of the areas that once experienced economic growth based on income from coffee that has gradually lost its value. Proper understanding of livelihoods, from the perspective of the people, is needed when policies and interventions are designed to react to these trends and to support rural people in a specific area. Insight is needed into the most important changes in the farming system, the most serious current problems in the farming system and off-farm activities, as perceived by farmers, and the strategies currently employed by the farmers to cope with these challenges. In order to understand the dynamics and interrelationships of livelihoods, a conceptual framework is needed that guides the data collection process and analysis.

The Sustainable Livelihoods (SL) framework (Carswell 1997; Carney 1998; DFID 2001) builds on the long history of rural studies making use of the concepts and elements of the 'peasant studies' of the 1960's and the multiple micro-economic models of the 1960s, 1970s and 1980s (Becker 1965; Barnum and Squire 1979; Deere and de Janvry; Sen 1981; Start and Johnson 2004). When Chambers (1987) and Chambers and Conway (1992) developed the concept of sustainable livelihood securities the field took a great leap forward in understanding the dynamics of rural livelihoods. Focussing on the prevailing interest in food security and the new interest in sustainability, they constructed a model on the basis of flows from stocks and capabilities through activities into well-being. At about the same time two other models were developed by Swift (1993) and Davies (1996). Swift emphasised the need to balance production and exchange with consumption. Central to this balancing were assets which consisted of investment, stores and claims, the latter of which was a non-material asset. Davies's model is very similar but uses Sen's (1981) entitlement concept. A few years later the current SL framework introduced five broad categories of capital, that is, human, social, natural, financial and physical capital. The older concept asset was used almost interchangeably with capital (Scoones 1998; Carney (ed), 1998; Moser 1998; Bebbington 1999; Start and Johnson 2004). The three stage micro-economic model (resources/assets, transformation, commodities) was still central. However, factors influencing the transformation were added.

'Livelihood' in the SL framework comprises a system made of *assets/capital, strategies* and *outcomes*. Assets or capital (human, social, natural, financial and physical capital) are the basic livelihood building blocks, the capacity people use in various strategies when striving for the objectives or the livelihood outcomes. The amount and type of capital used in the process influences the amount of outcome. Livelihood outcomes can be reinvested into assets/capital. The process is affected or guided by influences from government and private sector structures, laws, policies, institutions (formal or informal), culture (listed in the frame within *Transforming structures and processes*), and people's preferences and priorities. Changes in livelihoods can come in the form of seasons, slow trends or sudden shocks. These are represented in the frame by the *Vulnerability context*. "A livelihood is sustainable if it can cope with, and recover from, stresses and shocks, maintain or enhance its capabilities and assets" (Carswell (1997).

The SL framework is a useful analytical tool. As Ashley (2000) writes, it has an instinctive appeal, it is useful in generating insights and recommendations, it synthesises perspectives of different disciplines and provides an explicit focus on what matters to poor people. The SL framework has been widely adopted for structuring qualitative livelihoods-related development research and especially as a framework guiding participatory planning of development interventions. However, collecting quantitative data has many advantages over relying entirely on qualitative information. Project planning often has limited time and resources to do large quantitative surveys. However, this should not prevent planners from collecting quantitative data using a feasible sample size. A quantitative survey of this kind is not meant to be the only information source used in project planning, but to support qualitative approaches and secondary data. There is now a considerable body of experience in combining qualitative and quantitative approaches for example in studying poverty (da Silva 2006). In addition, the distinction between the two approaches is not always clear-cut. For example, qualitative investigations can benefit from rigorous sampling procedures (Levy and Barahona 2003) and some qualitative information can be quantified.

Whatever method is used, a framework is needed. This chapter tests the feasibility and value of applying the SL framework to a quantitative survey from the Taita Hills which was part of a bigger study of interactions of livelihoods, land use and environments in the highlands of East Africa (Soini 2006). The SL framework was used as a structure of this study both in the quantitative data collection and the statistical analysis of interrelationships of livelihood assets and outcomes.

STUDY SITE

Taita Hills (3° 24' S, 38° 20'E) is an island of fertile mountain surrounded by the dry bushlands. It is situated about 150 km from the coast of Kenya immediately south of the main road from the port town of Mombasa to the capital, Nairobi. To capture the variability in livelihoods, the study area (Figure 1) was a 28 km long and 6.5 km wide transect across the hills.

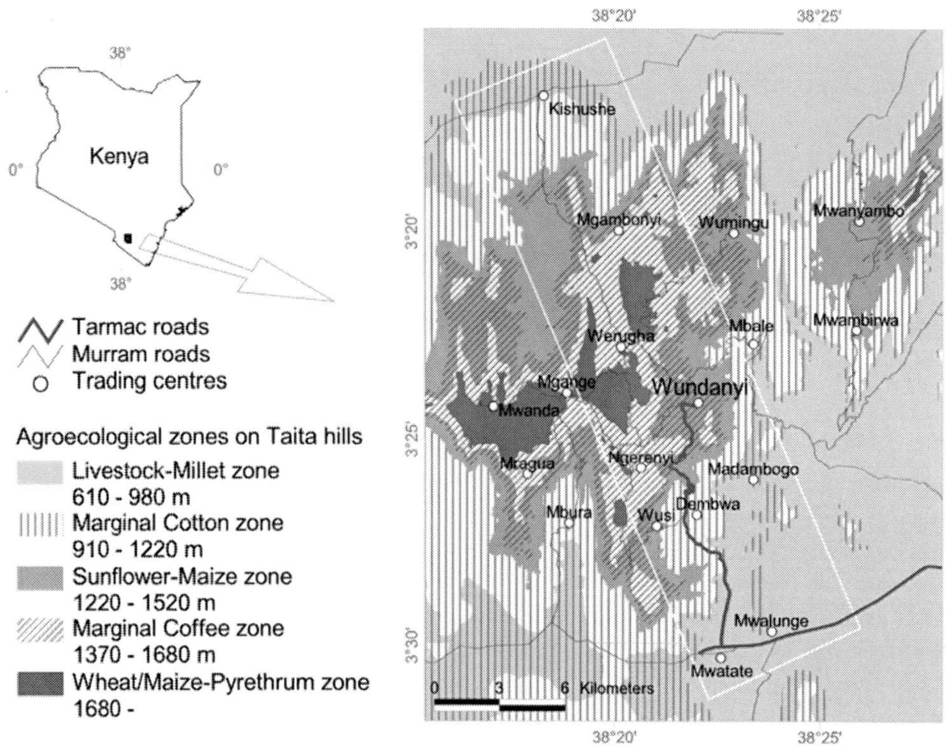

Figure 1. Study transect (in white) in the Taita Hills, Kenya. Agro-ecological zones drawn according to their altitudinal limits (Jaetzold and Schmidt 1983; Kenya 1989; Kenya 1991).

It reached from Mwatate, a trading centre on the plains (800m) on the southern side of the hills, across one of the highest peaks, Yale (2104m), to the dry and more remote northern plains (900m) at Kishushe. The biggest urban centre on the hills and Taita-Taveta district headquarters, Wundanyi, with a population of 4,000 (Kenya, 1997), was included in the transect. Five agroecological zones can be distinguished in the study transect, 1. (Midland) Livestock-Millet zone (790-980m), 2. Marginal cotton zone (910-1220m), 3. Sunflower-Maize zone (1220-1520m), 4. Marginal coffee zone (1370-1680m), and 5. Wheat/Maize-Pyrethrum zone (>1680m) (Figure 1). Annual rainfall during the two rainy seasons (March-May, September-October) varies from 480 to 700mm, 600 to 800mm, 700 to 900mm and 900 to 1200mm, and above 1200mm across the five zones respectively (Jaetzold and Schmidt 1983; Kenya 1989). Much less rain (250mm) falls on the surrounding plains.

Population density varies across the agroecological zones. The following figures were obtained calculating population density means in the study transect using Landscan (ORNL, 2000) data: 217 persons per sq km in the upper highland over 1680 m altitude, 121 in the Marginal coffee zone, 121 in the Sunflower-Maize zone, 44 in the Marginal-Cotton zone, and 50 persons per sq km in the Livestock-Millet zone. Population growth has been significant in the Taita Hills. The upward trend of population growth started in the mid 1920's.

By the end of 1960's population had almost tripled from the 40,000 in the early 1930's to about 111,000 in 1969 (Kenya Land Commission 1930, Census 1969, figures cited by Mkangi 1978). The district is now estimated to have slightly over 300,000 inhabitants out of which about 42,000 are estimated to reside in urban areas, mainly in Voi town on the plains (Kenya, 1997). Population pressure in the highlands has resulted in expansion of agriculture to the more marginal lowlands. More and more people move to the lowlands in search of land as large parts of the lowlands are not yet demarcated. According to a recent study about 64 % of the population in Wundanyi division and 58 % in Mwatate division live below the poverty line (CBS 2004).

METHODS

Fifty-one households were interviewed in 2004, 21 in the highlands, 15 in the northern slopes and lowlands and 15 in the southern slopes and lowlands. Grid sampling was used to select the households. Points one kilometre apart were drawn on the whole study transect and a random sample selected. A GPS was used to find the selected locations. When the GPS showed 0.5 km from the sample point the second household on the left was selected for the interview. This rule was used until about two thirds of the households were interviewed. Subsequently, the aim was to adjust the selection in order to have an equal number of old (above 50), middle-aged (36-50) and young (up to 35 years) heads of households from each of the three zones. This was to be done by finding the closest household to the one selected that represented the desired age group. However, it proved very difficult to find enough young families close to the sampling points in the highlands and the southern side of the hills.

Information was collected on the different livelihood assets: human (family structure, education, occupation), social (collective action, access to knowledge, links to sources of income outside the farm[1]), natural (land, livestock, trees and crops grown, access to common resources), financial capital (markets, off-farm jobs), and physical capital (proximity to roads, proximity to water source). Temporal change was emphasized. Perceived problems were documented. To measure priorities and values related to livelihood objectives, farmers were asked to identify the most desirable livelihood objectives.

To study interrelationships of livelihood capital and livelihood outcomes several variations of multivariate regression analysis were used. The framework does not lead into unique definitions of the indicators and decisions needed to be taken based on descriptive summaries of the data. An indicator representing livelihood outcomes was calculated as the mean of six components that were selected on the basis of desired livelihood objectives as perceived by the interviewed farmers (Table 6). All the components were calibrated to be either 0 or 1, or between 0 and 1, and consequently the indicator ranged between 0 and 1.

[1] It is due to the crucial requirement of the social linkage that contributions coming from outside farm are counted as social capital rather than financial capital. This follows the same principal as group membership being counted as social capital, even though the benefit of group membership is mainly financial.

Table 1. Independent variables used in the multivariate analysis. Each component scored either 0 or 1 or was an index defined between 0 and 1. k=upper limit of variable for which I=1 (see text)

Natural capital		Human capital		Financial capital		Social capital		Physical capital	
Component	Scores or k.	Component	Scores	Component	Scores	Component	Scores	Component	Scores
Total acreage	k = 5 ha	Labour force: (Number of dependants ≥ 10 y old living at home)	k = 7	Mother's off-farm job	1 = yes 0 = no	Outside contribution	1 = yes 0 = no	Proximity to water source	0 = >6 km 1 = < 1 km From 1km to 6 km, index decreases linearly
Inherited acreage	k = 5 ha	Young farmers	0 at 15 years 1 at > 40 years	Father's off-farm job	1 = yes 0 = no	Mother in a group	k = 3	Proximity to roads	1 = < 1 km to class 1 or class 2 road 0.5 < 1 km to class 3 road or < 2.5 km from tarmac 0 = > 1 km from class 3 road & on northern slopes
Improved cow	1 at > 1 cow 0.75 at one cow	Old farmers	1 at < 40 0 at 70	Children's off-farm jobs	k = 3 jobs	Father in a group	k = 3		
Number of hens	k = 40 hens	Fathers and mothers school years	k = 22						
Number of animals	k = 10 animals								
Plots in two ecological zones	1 = yes 0 = no								
Cash crops (Number of crops grown mainly for sale)	0.5 at one 0.75 at two 1 at > two cash crops								
Number of crops grown	k = 10 crops								

The components measured *housing* (stone 1 or mud house 0), *children's schooling* (1 at ≥ 5 children in secondary school), whether the farmer has *bought land* (1 yes, 0 no), *introduced animals* (1 yes, 0 no) or *crops* (1 yes, 0 no), or *established a business* of his/her own (1 yes, 0 no). The indicator representing livelihood outcomes was used as the dependent variable in the analysis. Independent variables represented five livelihood capital - human, social, natural, physical and financial. Indices (0 or 1, 0 to 1) for each capital were constructed by measurable components of each type of capital (Table 1). As the sample showed very skewed distributions of many of these variables, indices, I, were calculated on the 0-1 scale as

$I = x/k$ for $x<k$ $=1$ for $x>k$

Here x is the measured amount and k a suitable upper limit, defined for each variable (Table 1).

RESULTS

Only the main findings of the livelihood survey are summarised in this chapter. A more detailed descriptive analysis of the survey is available in Soini 2005. This summary concentrates first on the trends and problems with all the capital categories (natural, financial, human, social and physical capital) and looks at the different coping strategies farmers use during sudden shocks. Subsequently the results of the study on interrelationships between livelihood capital, strategies and outcomes are presented. Table 2 summarises basic farm and household characterisation information.

Table 2. Some important farmer and farm characteristics in the study transect in the Taita Hills, from the interview survey of 51 households

Socio-economic indicators	Frequency
Average number of years at school	6.8 fathers, 5.5 mothers
More than 8 years at school	32 % fathers, 10 % mothers
Vocational training	61 % fathers, 25 % mothers
Off-farm job (both casual and permanent)	50 % (fathers), 19 % (mothers)
Farmers belonging to groups	39 % fathers, 54 % mothers
Average number of persons living in a household	5.5
Average family size	6.5
Grandparents/grandchildren or others living in a household	53% of the households
Highland farmers who have a lowland plot	10/21
Average number of plots per family	2.1
Average plot size	2.3 ha
Farmers who have inherited/been allocated (by gov.) land	Inherited 84 % / allocated 27 %

Table 2. (Continued)

Socio-economic indicators	Frequency
Farmers who have sold land	8 %
Farmers who have bought land	27 %
Farmers who rent land	6 %
Average distance between different plots	1 hour (about 5km)
Average total plot area per family	4.6 ha (range 0.4 –26 ha)
Farmers living in a stone house/mud house	29% / 71 %

MAIN TRENDS AND CHANGES IN LIVELIHOOD CAPITAL

Table 3 lists the trends that have happened within one generation as identified by farmers as the most conspicuous affecting the Taita Hills. Those that score high are all related to natural capital. Table 4 lists the main problems encountered in the farming system, as perceived by farmers. Again problems related to natural capital score high, however, lack of financial capital for farm investments is amongst the most pressing problems. Soil depletion is generally perceived by the Taita farmers as the most conspicuous change affecting cropping and resulting in more work and demand of farm inputs. A majority of farmers perceive a decrease in yields of all the main crops. The decrease was always related to decreased soil fertility. Also, pests, mainly Larger Grain Borer, *Protephanus trancutus,* currently destroy a large part of the maize harvest in the Taita Hills. Related to the productivity and pest problem (natural capital) is the problem of inadequate financial capital to buy farm inputs (pesticides, fertilisers, cow dung etc).

None of the farmers mentioned land consolidation amongst the commonly recognised trends. Contrary to expectations, the agricultural land reform (Swynnerton 1954) of land adjudication, consolidation and registration which has been partly implemented in the Taita Hills, land consolidation has obviously not happened to that extent that it would have completely disrupted the utilisation of different ecological zones by the highland farmer families. In some areas farmers have, in fact, refused land consolidation as they have found the exercise too cumbersome, involving displacement of people and creating conflicts. Taita farming system has traditionally made use of the varying ecological conditions of the different altitudes of the hills and adjacent plains (Nazzaro, 1974; Fleuret, 1988). Planting crops both in the plains and hills, a family can more nearly assure themselves of a dependable food supply from season to season. As it is still possible to buy plots in the lowlands to supplement a highland plot, highland farmers are in a better position to maintain the security of different agro-ecological zones. Half of the highland farmers have lowland plots. Most (70%) of the bought plots of this study were in the lowlands. It is much more difficult to get hold of a highland plot. Due to population pressure in the highlands an increasing number of people are forced to depend on lowland plots only.

Already in 1926 a 'Native forestry instructor' was employed by the Native Council to address the serious problem of deforestation. However, the perception of whether there are fewer or more trees or tree species at present than before, varies from one farmer to another.

When answering the questions, farmers were mainly assessing the trend in their own neighbourhood.

Table 3. Perceived trends since fathers' time (since the interviewee was living in his/her father's household) in the Taita Hills. Frequency indicates how many farmers mentioned a trend (n=51).

Category	What is different now compared to when your father was a farmer?	Freq
Cultivation	One needs to use manure, fertiliser, (compost)	24
	Use of spacing in cultivating	14
	Soils have depleted	9
	Using deep tillage	8
	Use of certified seeds	7
	More crop species	6
	Use of pesticides	5
	We have improved knowledge to cultivate and keep livestock	5
	Bigger area of land under cultivation	5
	Soil conservation (bunds)	5
	Intercropping	3
	Timely planting	3
	Agroforestry	3
Trees	There are fewer trees/more trees	27/10
	There are fewer tree species/more tree species	15/13
	New tree species like Grevillea, Avocado, Cypress	6
Animal rearing	Animals are kept in zero-grazing	20
	Farmers keep fewer animals	16
	Planting/cutting grass for animals	13
	We obtain more milk	9
	Improved animals	8
	Animals in the farm not group grazing	7
	Poultry keeping	4
	Dipping or spraying of animals	4
	More manure obtained from zero-grazing	4
	De-worming of animals	4
	Artificial Insemination	2
Other	Off-farm cash income source	11
	We cultivate vegetables	6
	Rains have decreased	4

Table 4. The most common problems perceived by farmers in the Taita Hills. Frequency indicates how many farmers mentioned a trend (n=51).

Problems with crop growing	Freq
Pests (Larger Grain Borer, *Protephanus trancutus*)	37 (10)
Lack of capital to invest in farm inputs	33
Drought	29
Animal destruction (baboons, monkeys, but elephants in 11 cases in the lowlands)	23
Extension is not available[1]	11
Shortage of labour	10
Theft / flooding / moles	3 (each mentioned only once)
Problems with tree growing	
Unavailability of seedlings (especially grafted)	42 (23)
Drying	5
Too small a plot / pests on fruit trees / seedlings expensive	3 (each mentioned only once)
Problems with livestock	
Diseases (tickborne, newcastle in chicken, foot and mouth, east coast fever, ccpp)	36
Lack of enough fodder esp. dry periods	20
Expensive drugs and vet services	10
No dipping facilities	6
Problems with off-farm activities	
Lack of capital to keep good stocks	7
Lack of enough customers	6
Getting permits	2
Lack of timber for furniture making / no good equipment / customers do not pay / transportation	4 (each mentioned only once)

[1]The question on extension service availability was not mentioned amongst the most important problems. It came up when asking about the availability and access to farm inputs and knowledge. All the rest (40) say it is available, but only on request. 41 farmers say veterinary services are available on request, one mentions that vet people even travel around on farms.

It seems that some neighbourhoods have planted more trees and introduced new species while some other neighbourhoods have done the opposite. There is no significant association between farmer's age and whether s/he perceives an increase or a decrease in the number of trees. This suggests that the difference is mainly spatial. Those farmers who perceive an increase are in the highlands or in the northern plains. Sixty-one percent of the interviewed say they have enough firewood from their own land. Crop residues, particularly maize cobs and stalks, are also used as fuel. Twenty-seven percent of the farmers say they have enough trees to meet their timber needs. Most of the farmers who reported of having enough trees on their own land to meet their firewood or timber needs live in the highlands or on the northern side of the hills.

There have been many tree planting programmes over the years in the Taita Hills, but farmers perceive unavailability of seedlings as the most important constraints to tree planting. This is contrary to the common belief that it is unavailability of land that stops farmers from growing more trees.

Due to agricultural intensification, livestock numbers have gone steadily down and most of the animals in the highlands are now kept in zero-grazing. Practically all farmers who keep improved cows (in the highlands) cultivate fodder grass(es) on their land, and most of them have enough fodder from home. Farmers estimated the average number of animals their fathers used to have as 11 cows and 13.9 goats. In the highlands it is now typical for a farm to have one or two improved cows. Lowland farmers have typically 3-4 mixed or zebu cows. The average number of goats or sheep per household is 7.4. Most of the farmers had had their largest number of animals 5 to 10 years ago. Lack of funds to buy veterinary services (like any other farm inputs) is the underlying cause of the most serious problem for livestock keeping: livestock diseases.

A lot of changes and adjustments are often made on farms. These are not the same as the perceived common trends (Table 3), but are farm specific changes that each farmer reported. Almost half of the farmers (21/51) had introduced some new crops on their farms between 1982 and 2003 (mean 1996). However, it is not possible to identify typical choices. They include tomatoes, onions, groundnuts, beans, cowpeas, cotton, green grams, potatoes and many others. Twenty-eight farmers mention spacing as a newly introduced way of cultivation (introduced between 1983 and 2002). Fifteen said they have started to add manure or fertiliser (1978-2003), seven farmers have introduced pesticide use (1990-2003), six have constructed bunds (1985-1999), five introduced deep tillage (1998-2003) and four started to harvest rainwater for irrigation (2000-2003). Forty out of 51 farmers mentioned some changes in livestock rearing, the most typical new rearing technology being zero grazing (19 farmers) (introduced between 1974 and 2004). Others include de-worming (8), introduction of dairy cows instead of Zebus (6), a dairy goat (1), dipping cows (6), and fodder production (2). Amongst the trees more often introduced on farms, *Grevillea* scored highest (18) followed by *Mangifera indica* (16) and *Persea americana* (11).

About half of the farmers have abandoned some activities or technologies they previously used. The abandoned technologies and activities represent again a very large variety, including cultivation of tomatoes, keeping rabbits, using fertiliser or manure, keeping poultry and sheep, cutting all the *Cupressus* trees etc. Looking at the introduced and abandoned farm activities or technologies no trends emerge and no shift in the farming system as a whole seems to have occurred. When one farmer introduces tomatoes another abandons the crop, when one starts using fertilisers and manure another farmer abandons the practice. If any trends can be found from within these farm specific changes, it is the replacing of indigenous trees with fewer exotic species, spacing used in planting maize and beans, and zero grazing of livestock, and increased usage of fertiliser or manure. None of the farmers mentioned the decreasing coffee price as a problem or a trend. In fact only one farmer cultivating coffee was captured by the survey, another had abandoned coffee growing due to non-existent markets. The District development plans (Kenya 1976; 1989; 1997) have continued mentioning the importance of coffee amongst the cash crops in the district.

Most of the farmers cultivate mainly for home consumption. However, selling even a small proportion of their crop production is an important cash source. Table 5 lists the main crops grown by the farmers interviewed and the proportion used or sold. In addition to crop sales, more than half (53%) of the interviewed farmers sell some tree products, the most typical being fruits (33 %). Twenty-two percent of the farmers sell timber, 16 % sell firewood (one of which sells charcoal), 6 % sell nuts, and some few sell fodder and poles. *Mangifera indica, Persea Americana,Carica papaya* and *Psidium guajava* are clearly the most important trees to yield fruits for sale. Top five species for timber and firewood sales were *Grevillea robusta, Cupressus lusitanica, Eucalyptus* ssp., *Melia volkensii* and *Acacia* sp.

Table 5. Main crops grown by the interviewed households (n=51) in the Taita Hills, and proportion of home usage and selling. Note that coffee is no more an important cash crop in the Taita Hills.

	Number of farmers who...				Total number of farmers growing
Crop	use all the produce at home	use more at home than sell	use half and sell half	sell more than they use at home	
Maize (*Zea mays*)	42	8		1	51
Beans (*Phaseolus vulgaris*)	37	8		4	50
Cassava (*Manihot esculenta*)	22	2	3	4	32
Cowpeas (*Vigna unguiculata*)	22	3		1	28
Mango (*Mangifera indica*)	7	6		1	14
Pigeon peas (*Cajanus cajan*)	11	1		2	14
Banana (*Musa* spp.)	8	4		1	13
Millet (*Eleusine coracana*)	11	1			12
Green grams (*Phaseolus mungo*)	6	4		1	11
Cabbage (*Brassica oleracea*)	1	1		6	8
Irish potato (*Solanum tuberosum*)	4	1		3	8
Sweet potato (*Ipomoea batatas*)	6	1			7
Tomato (*Lycopersicon esculentum*)	6	1			7

Other crops grown by fewer farmers in the sample include arrow root (*Canna edulis*), kales (*Spinacea oleracea*), beetroot (*Beta vulgaris*), cauliflower (*Brassica oleracea*), coffee (*Coffea arabica*), cotton (*Gossypium* spp.), groundnuts (*Arachis hypogeoa*), hot pepper (*Capsicum* spp.), lettuce (*Lactuca sativa*), *Catha edulis*, onion (*Allium cepa*), pumpkin (*Cucurbita pepo*), sweet pepper (*Capsicum annuum*), spinach (*Spinacia-olerecea*), sugar cane (*Saccharum officinarum*), sunflower (*Helianthus annuus*).

However, the largest proportion of any tree product is used at home. Poultry sale for cash plays a very important role in Taita households. Nearly all households keep hens, some of them as many as 90. Naturally, for those who keep large numbers it is a family business. Both eggs and broilers are sold. Milk is another very important income source to many (33%) families. All except one of those who sell milk consider it an important or very important addition to the household income. One to ten litres are sold per day (mean 4.5 l), which makes between 20 and 240 KShs per day (1 $ = 75 KShs).

The increasing need of cash for farm inputs, school fees, healthcare, combined with the decreasing farm produce has forced farmers to look for off-farm jobs. Many more fathers (50%) than mothers (19%) have an off-farm job. But the present options for off-farm income are seriously limited and the capital to set up businesses inadequate. There are also too many businesses of the same type leading to lack of enough customers.

Inadequate capital to buy farm inputs (pesticides, fertilisers, cow dung etc) was seen as one of the biggest problems and depleted soil fertility as the most conspicuous trend. It is interesting, however, that the survey data does not show any correlation between fertiliser use and whether the farmer has an off-farm income. The income that is derived from an off-farm job is used for something more pressing or less risky than farm inputs such as fertiliser. Often periods of heavy cash demand and availability do not match within a year (69% percent of the farmers reported). Three quarters of the farmers cannot either get credit from anywhere.

There was also no correlation between fathers' educational level and whether he has a job. This is due to the fact that most of the jobs available do not require formal training. Fathers' vocational training was usually in the fields of building (carpentry, masonry, painting etc), driving and mechanics, but there was a large range of other fields as well such as teaching, hotel keeping, agriculture, police, store keeping, theology, sign painting, office administration, cooking, army training, finance or civil engineering. Mothers' vocational training represented much less variation the majority being in tailoring. Others mentioned included clerk, typist, pot making, teaching and forestry. While ten of the fathers (n=51) had had an opportunity to finish secondary school and two of them had even continued further, only two of the mothers had finished secondary education (11 years). There is no clear correlation between the age and number of school years in either of the gender groups. Obviously young, more educated people have not stayed as farmers.

More than half (27/51) of the households have grandparents, grandchildren and others living on the farm. These can be considered both as labour force as well as additional consumers of the farm produce. However, one fifth of the farmers, particularly lowland farmers, consider lack of labour as a serious limiting factor to their farming activity and as a consequence part of their farmland is left uncultivated.

Taita households are very actively involved in group activities. Most of the groups are women's or village merry-go-rounds (groups that save collectively to one member at a time), but there are also a few specialised project-based groups. All, except one of the women and four of the men belonging to groups, mentioned some benefits. In most cases it is money to buy household assets or pay school fees. Some mention spiritual support and new ideas from the group as benefits. One fourth of the group members think they can obtain credit from their groups, however, noone had actually tried. In addition to group activities i.e., organized collective action, most of the households (41/51) help each other with things like firewood, water, food, labour, cash by borrowing, and lending and borrowing household assets like wheelbarrows, bicycles, kitchen utensils etc.

One fifth of the households receive financial contributions from someone outside the farm. These are most often adult children, but can also be other relatives. But interestingly, a higher percentage (35%) of households contribute in some way to the well-being of someone outside the farm. These beneficiaries are most often grandchildren, parents or other members of the extended family.

There is large variation in access to a water source between the households with nine households out of 51 having a water tap within the homestead and the longest distance to water being 12km. Fifteen households fetch their water from a river, three from a spring, one from a tank and the rest from a tap (either communal or their own). Only two homesteads have irrigation water available.

There is also clear differentiation of households with access to transport. Wundanyi, the District headquarters on the hills is reached from Mwatate on the Taveta-Voi road on the southern side of the hills by a tarmac road (class 1). Tarmac road leads all the way to Voi town on the Nairobi-Mombasa road. Best access to markets and services outside the Taita Hills are naturally along this road. It is easy to get public transport up and down the hill along this road. Most of the other roads on the Hills may be even impassable during part of the rainy season. The northern side of the hills is furthest from markets, except small local ones like that of Kishushe itself, and least accessible.

Coping with Shocks

When a sudden need of cash arises in the village or at home due to a funeral, sickness or a wedding, strategies are needed to cope with the additional expenditures. Very many different coping strategies and combinations of these were mentioned by the farmers, including selling an animal, selling chickens, selling timber, help received from the family (children), relying on shop/kiosk/handicrafts, selling crops, casual work, borrowing money, selling milk, selling eggs, selling fruits, finding gemstones, selling charcoal, squeezing the family budget, using previous harvest, and relying on village collection.

Understanding coping with shocks can be based on considering assets, activities and needs or preferences (Devereux 1993; Start and Johnson 2004). Coping at *asset level* involves liquidating stores, eroding assets, or calling down informal or formal claims. Coping at *preference, need or consumption level* typically involves consumption diversification, rationing or reduction. Coping at *activity level* can involve working harder or longer hours, or diversifying into new activities. Sixty-five percent of the Taita farmers can cope through shocks by coping at the needs/preference level or activity level and not depleting any of their assets. However, one third of the farmers had to sell an animal in order to cope with a sudden need at home, a response at the asset level.

Village collection (calling down informal or formal claims) is an important way for farmers to help each other in coping with sudden or emergency cash needs. But only 14% of the farmers say they have been able to count on collection during times of additional sudden expenditure. Taking part in village collection to help someone else is easier to handle. Most often squeezing the family budget in sufficient, or one might sell more crops or a hen.

To cope through a drought, 35 % percent of farmers mention they had to sell animals, goats or cows. Fifty-nine percent of the farmers could cope with the last drought by means that did not deplete the asset base and usually by one or two means only. Selling land is probably the most extreme method that a farmer needs to use in order to cope with his/her cash needs. Only four farmers out of 51 admitted they had sold some of their land.

Interrelationships between Livelihood Capital, Strategies and Outcomes

How do farmers themselves think they could cope better and improve their livelihoods? Table 6 is a list of answers to this question of livelihood objectives or desired material wealth in life in the Taita Hills.

Table 6. Livelihood objectives in the Taita Hills as perceived by farmers (n=51).

Livelihood objective	Freq
Construct a good and big stone house	37
Acquire a dairy cow or goat or expand dairy farming	31
Establish a kiosk or a shop	29
Expand or start poultry	25
Invest to a water tank	17
Have electricity to their house, 5 of which mention specifically solar power	14
Acquire more animals	12
Arrange irrigation to the farm	11
Expand land area	10
Build terraces	9
Start or expand horticulture	9
Buy farm inputs like fertilisers, manure, pesticides	8
Plant fruit trees	7
Improve current business	7
To have water pumped to compound, some especially for irrigation purposes	6
TV, furniture, tractor, plough, mill and rental houses	
Bakery, grocery, tailoring, timber sales, rental houses	
Beekeeping, groundnuts, proper planning, timber trees	

The livelihood objectives that were considered by the farmers themselves as the most desirable in improving the standard of living were used to select the components of the livelihood outcome index. The outcome index, which is the average of the six components, varies between 0 and 0.67. All the components are roughly equally prevalent (means for 5 of the components range from 0.22 to 0.29), with *introduced crops* (0.39) slightly more so. The two components, *bought land* and *children's schooling* are significantly positively correlated. This is probably due to the fact that older couples more often have secondary school aged children and due to their age they have in general had more opportunities to invest and buy assets. However, the general lack of correlation between the livelihood outcome components suggests that there is no consistent pattern in achieving outcomes – people do not consistently get one, then the next.

Principle component analysis on the livelihood outcome components (not shown) was used to study correlations between the different outcome components. *Bought land*, *established business* and *introduced crops* are as close to unrelated as possible. The analysis did reveal two obvious groups of households, those with and without introduced crops. Otherwise farms do not fall into distinct groups defined by patterns of outcomes. However, when the data from farms in the highlands, southern side and northern side of the hills are considered then a clear differentiation can be seen in outcome scores (Table 7). Highlands and the southern slopes and plains are better off on all outcomes except crop introduction.

Table 7. Average scores in outcome index components in the northern and southern slopes and the highlands of Taita Hills.

	Average of					
Zone	Housing	Children in Secondary school	Bought land	Introduced animals	Introduced crops	Established business
North	0.13	0.20	0.13	0.13	0.47	0.17
South	0.33	0.23	0.20	0.27	0.33	0.27
Highlands	0.38	0.45	0.43	0.24	0.38	0.24
Grand Total	0.29	0.31	0.27	0.22	0.39	0.23

Although the process is dynamic, with each household following a different time trajectory, it is expected that some links between the assets/capital and the outcomes can be detected in cross-sectional data. A linear regression model, with the livelihood outcome related to the five livelihood capital indices, is an approximation used to explore possible correlations.

Results of the linear multivariate regression analysis are shown in Table 8. Only financial capital has a clear, positive relation (significant at $p=<0.01$) with the outcome. In addition, natural capital has a positive relation, but it is not statistically significant. Human, physical and social capital show no relationship with the livelihood outcome. The relationship with financial capital is different in the three zones (not shown), with the importance of financial capital being clearest in the highlands.

It is important to note that the variance accounted for is small. Most of the variation in the outcome index is simply not linearly accounted for by the four assets. Several explanations for this are explored below.

First, one can claim that the simple linear model is unrealistic, particularly as the outcome is constrained between 0 and 1. A logistic model was also tested. However, the result using this model was almost exactly the same. Financial capital is still the only one clearly important.

Second, it is recognised that the sustainable livelihoods theory claims that all the capitals are needed for successful livelihood outcomes. However, the two previous models assume that all five have independent influences on the outcome. So a new variable *All capital* was added. It is the product of all the five capitals and reflects the overall interaction between them, being zero if any one component is zero. However, this does not make any difference to the results, with this interaction variable being unimportant in explaining variation in outcome index.

Table 8. Results of the linear multivariate regression analysis aiming at detecting relationships between four groups of livelihood capital (indices) and livelihood outcomes as an index measuring desired outcomes as perceived by the interviewed farmers in the Taita Hills.

Parameter	estimate	s.e.	t(45)	t pr.
Constant	0.008	0.100	0.08	0.936
Natural capital	0.222	0.151	1.47	0.148
Financial capital	0.435	0.119	3.65	<.001
Social capital	-0.109	0.149	-0.73	0.470
Human capital	0.143	0.138	1.04	0.305
Physical capital	0.1142	0.0893	1.28	0.208
Percent variance accounted for = r^2 = 26 %				

Third, it may be that the influences of the assets are different for each outcome, so no clear results are seen when we analyse the average outcome. This is investigated by analysing each of the six components of the livelihood outcome separately. As these have many 0 and 1 values, the logistic model is used. Some associations show up. *Introduced animals* is positively associated with natural and physical capital. *Introduced animals* is, however, negatively associated with social capital. One may argue that people who join groups, or rely on outside contributions to start with (the indicators of social capital) are those with no opportunities to get animals, or further, that farmers with animals are too busy to take part in group activities. As expected *Established business* shows strong association with financial capital. It also shows negative association with social capital. Farmers with no business may be more often members of groups as they depend on group collections as their coping strategy. *Children's schooling* (number of children in secondary school) is positively correlated with natural capital. This may indicate that families who have older children, have also had more time to invest in natural capital. *Introduced crops* is marginally negatively associated with financial capital. The association is not statistically significant (0.182). However, it may show that having a job leaves limited time for labour intensive activities like new crop enterprises. The case of fertiliser use earlier showed that off-farm income is not invested into farming activities.

Fourth, a further explanation is simply that there are many individual livelihood strategies, or that the framework does not capture the important aspects or explain the linkages, and it is not possible to pull out any general patterns (at least in a sample of 51). The logistic model shows the best relationship, but some households are far off what could be expected. There are households with a score of 0 in all the livelihood outcomes, yet they can be reasonably endowed with some of the capitals. There are also some farms with high scores on outcomes, yet some of their capital scores are 0.

Further explanations were sought by replacing the five capital indices with their components. A few modifications were made: *Inherited acreage* was dropped as it is highly correlated with *Total acreage*. *Number of improved cows* and *Number of hens* were also dropped as these are almost the same as *Introduced animals* which is one of the components of the livelihood outcome index, and *Number of animals* still remains to measure livestock influence. Father's and mother's group membership were merged as they are highly correlated (either both men and women or neither tend to be in groups, with a few men in groups on their own.) Table 9 shows the results of the analysis.

Table 9. Regression analysis of associations of livelihood outcome (index) with livelihood capitals replacing the four capital indices with their components.

Parameter	estimate	s.e.	t(34)	t pr.
Constant	-0.264	0.262	-1.01	0.319
Total acreage	0.0250	0.0876	0.29	0.777
Number of animals	0.1655	0.0842	1.97	0.058
Plots in two ecological zones	0.1012	0.0686	1.48	0.149
Cash crops	-0.0044	0.0720	-0.06	0.952
Number of crops	-0.125	0.165	-0.76	0.454
Labour force	-0.007	0.105	-0.06	0.950
Young farmer	0.330	0.159	2.08	0.045
Old farmer	0.083	0.159	0.53	0.602
Father's and mother's school years	0.132	0.106	1.25	0.221
Outside contribution	-0.0619	0.0760	-0.81	0.421
Father and mother in groups	-0.178	0.175	-1.02	0.315
Mother's off-farm job	0.1448	0.0701	2.06	0.047
Father's off-farm job	0.1734	0.0707	2.45	0.019
Children's off-farm jobs	0.034	0.114	0.30	0.765
Proximity to a water source	-0.0450	0.0844	-0.53	0.597
Proximity to roads	0.2035	0.0852	2.39	0.023

It is surprising that Total acreage is not correlated with the livelihood outcome index. However, in Taita Hills many farmers mentioned that they cannot cultivate all of their land due to shortage of labour. The off-farm income variables (Mother's off-farm job, Fathers off-farm job) are strongly correlated with overall outcomes showing the significance of financial capital in achieving livelihood outcomes. Number of animals also contributes clearly to the livelihood outcome. As the index Young farmer is small for young farmers and increases to 1 at age 40, the positive relationship indicates that older farmers up to age 40 have a higher score on the outcome index. This is reasonable if people take some time to acquire assets and outcomes. The trend does not continue after the age of forty (Old farmer does not show an association). Proximity to roads is positively associated with achieving livelihood outcomes. Having plots in two ecological zones has a marginal positive association.

Discussion

The SL framework which has been found useful in structuring qualitative investigation, is also useful in the design of quantitative surveys. Using the SL framework focuses the survey instrument on the key elements of livelihoods. It also helps structure the presentation and analysis of the data. However, some problems arise from working with the SL framework.

Livelihoods are presented in the frame as a *system* consisting of assets/capital, strategies, outcomes, transforming structures and processes, and vulnerability. In the simplified framework vulnerability context has been dropped, and transforming structures and processes are named as external influences. People's preferences are placed together with these external influences in the middle of the framework. What is external and what is internal in the livelihoods (system) is problematic. Natural capital is understood as an internal component, however, one could argue that natural capital is both the natural resources utilised (internal) and the environment, the setting of livelihoods (external)(see Soini 2006: 4). This raises the question of where to set the physical or geographical boundary of a person's livelihood in the framework.

One of the main difficulties is related to the selection of the indicators of both assets and outcomes. If the researcher determines outcome definitions then there is a danger that they represent researcher's (outsider's) livelihood objectives and not the livelihood objectives of the local people. The livelihood outcomes listed within the framework (empowerment, more sustainable use of NR base, increased well-being etc) are not easy to capture in a quantitative form. More importantly they are usually not things that farmers specifically mention as desired livelihood outcomes. But farmers' perceived livelihood objectives as presented in the data are not easily used as livelihood outcome indicators either. A wide range of farmer objectives can be included in the concept of 'well-being', but again decisions to do so are the researcher's, not farmer's. Quantitative methods have been criticised of being researcher centred. In this case, however, this study particularly opted to use farmers' own stated livelihood objectives as components of the outcome index. These objectives were obtained in the interviews by asking what would be the most desirable 'things' (desired state, assets or activities) that the household strive to attain and which would increase the standard (state, 'hali' in Swahili) of living. The answers were from the individual household point of view. Many of the things farmers listed belong more clearly to the category of assets or strategies than to the category of outcomes. The fuzzy line between assets, strategies and outcomes may be one of the main difficulties in operationalising the framework. This is obvious from the framework as assets lead to outcomes and outcomes feed back to assets. This constant feedback loop in time makes it difficult to capture what comes first and what follows it. For example, to get a proper stone house is a typical farmer's livelihood objective. A stone house is an asset, but it is difficult to separate it from the outcomes it brings in the form of comfort, esteem, happiness, pleasure of beauty, ease of maintenance. When a stone house, an asset, is obtained, it immediately leads to some livelihood outcomes without any transformation through strategies (activities). In another example, farmers commonly thought that to introduce crops or animals (principally a dairy cow) would make a big improvement to the standard of living. Here the farmer's objective is to employ a strategy or an activity (growing vegetables, keeping a cow) or to own an asset (e.g. a cow). A cow as an asset, just like a stone house, can be an outcome directly in the form of prestige or self-esteem, and it leads to an outcome of improved children's well-being and health through simple activities of milking, boiling the milk and giving the milk to the children. Further, some of these direct outcomes like prestige and self-esteem are the foundations of empowerment. This difficulty arises from trying to use cross-sectional data collected at a single time point to understand dynamic processes. Although farmers' recall and perceptions of trends are collected, it is hard to see how to build these into quantitative analyses described by the framework.

Several attempts have been recently made to improve the framework. These ideas have remained mainly in unpublished papers or working paper series. Additional elements have been introduced, such as personal assets (Hamilton-Peach and Townsley 2004), information capital (Odero 2003), cultural capital (Bebbington 1999) and gender, age, class (or caste) and ethnic group (as 'social' processes), technologies, and aspirations and opportunities (Hamilton-Peach and Townsley 2004; Start and Johnson 2004). Strategies have been renamed as actions (Hamilton-Peach and Townsley 2004). The 'PIP box' (policies, institutions and processes) has been unpacked (e.g. Hobley 2001; Hamilton-Peach and Townsley 2004) to allow more specific emphasis on the various contents (such as markets, politics, culture, rights, enabling agencies and service providers) within it. Linkages have been emphasised, and elements of the framework have been rearranged, for example making it less sequential and placing the poor in the centre (Hamilton-Peach and Townsley 2004).

One factor making the reality difficult to model is the apparent individuality of farmer decision-making. For example, the study showed that families are often shifting activities without an obvious trend within the community. When one family starts growing vegetables, another family stops growing them, when one starts using fertilisers, another stops using fertiliser. It was not possible to find tangible reasons (like markets, projects or promotion of a technology by extension) for this. This clearly makes it difficult to capture cause-effect relations. In the context of inadequate information on the technology and the markets, much of this alteration of the farming is based on 'trying one's luck' ('kubahatisha' in Swahili) with different alternatives that are readily at hand. Capturing reasons for differing farmer behaviour is one of the strengths of qualitative investigations.

Adoption studies have added some understanding of the decision-making aspects of selecting household strategies. They have mainly focussed on technologies introduced by development interventions, but the same mechanisms of decision-making are at play independently from the source of the new technology. Pattanayak *et al.* (2003), by reviewing 20 articles on adoption of agricultural and forestry technology by smallholders concluded that it is the following five categories of factors that explain technology adoption: market incentives, risk and uncertainty, biophysical factors, resource endowments and preferences. Out of these categories, the livelihood framework captures risk and uncertainty in its vulnerability context, biophysical factors mainly in its natural and physical capital, resource endowments in its five capitals/assets, and preferences as preferences in the simplified SL framework. Markets are not mentioned specifically in the transforming structures and processes box of the SL framework nor are they present in the simplified version of the framework. However, markets can be part of the framework indirectly under financial capital (sale) and physical capital (presence of markets, infrastructure).

Preferences are clearly related to personal capital and they guide decision-making. Both concepts try to capture aspects such as personal aspirations for change, motivations, opinions and attitudes and the drive to assert personal rights. The desire and ability to engage in political activity and the spiritual side of life (Hamilton-Peach and Townsley 2004) can play a considerable part in guiding preferences. They also influence how people form perceptions of risk that guide the choice of a strategy. Another concept, capability (Sen 1985), describes the personal and often non-economic and non-social attributes of an individual. Mainstream definitions and measures of human capability often focus on health, nutrition and education and include the more obvious productive attributes, such as skills, strength, information and knowledge (Start and Johnson 2004). Human capital contributes to quality of life in many other ways than just by its impact on productivity, a range of ways that is summed up as human capability. This includes the ability to debate, negotiate, to add voice to political processes and, ultimately, to be an agent of social change. It defines the advantage of that person relative to someone else and hence influences the opportunities open to them (Start and Johnson 2004).

Recognising that potential opportunities are not identical for all (Barrett et al. 2006), and they come and go, opportunities have been suggested as a separate factor to be added to the livelihood framework (Hamilton-Peach and Townsley 2004). One person's opportunity is often another person's lost opportunity. Or, one man's social capital, is another's social disadvantage (Start and Johnson 2004). Most of the latent opportunities do not exist in isolation for one individual or a household. This is particularly true in times of crisis when competition between individuals and households is harder. As the study by Eriksen *et al.* (2005) concludes, social relations play a crucial role in which niche each individual and household end up occupying during crises like a drought. The SL framework does not in general capture this idea of personal livelihoods existing in the setting of other people's livelihoods. Social capital is generally understood as a social resource, not as a competing force.

A considerably larger sample size may allow detection of more correlations. However, it is likely that the links would still be weak with a lot of variation. Also, in practice, the sample size is often a compromise between availability of resources and the optimal quantity and quality of information needed to draw credible conclusions. Large levels of variation and weak correlations in the quantitative analysis could be seen as a weakness of the approach. However, there can be important inferences to be drawn from that variation.

The study highlights the high level of individual or household level variability (actual household level changes) that is different than the reported common trends (Table 3). There is great variation in what farmers report as changes, perceived problems, adaptation and coping strategies. There is also great variation in selection and experimenting with crops. Twenty-eight crops (excluding tree crops) were documented, but only four of them are grown by more than half of the farmers. There is also great variation in the success in achieving desired livelihood outcomes.

What can this variation mean in practice? One consequence of the diversity and continued switching between farm enterprises is the requirement for dynamic support from the district agricultural officers and projects. They would need to continually expand the range of information and support they give, rather than promoting single solutions. The useful approach for the extension service or projects will be to continually increase the options. Also, development projects need to understand that households are increasingly multi-occupational. In addition, the portfolio of occupations changes constantly while farmers adjust their strategies. This has consequences especially for time available for activities introduced by a development intervention. The shift from agriculture to non-agriculture earning is happening in many rural areas in Africa (Livingstone 1991; Haggblade *et al.* 1989; Reardon 1997; Bryceson 1999; Soini 2005). Even though most of the people living in rural areas consider themselves farmers, their livelihoods are increasingly multi-occupational.

Conclusion

Rural agricultural livelihoods and the environment are highly interdependent. Farming systems drive environmental change and the ensuing environmental change drives the developments of the farming system. There are some trends that are commonly perceived as changes in the environment and the farming system. Some of the biggest changes are declining soil fertility and decreasing crop yields, decreasing number of livestock due to decreasing farm size, new technologies such as spacing and zero-grazing, changes in the number of trees and tree species (perceptions vary), and the rising need to have off-farm activities to supplement farm income. There is, however, great variation on what farmers report as changes on their farms, perceived problems, adaptation strategies to changes and coping strategies. There is also great variation in selection and experimenting with crops. Twenty-eight crops (excluding tree crops) were documented in this study, but only four of them are grown by more than half of the farmers. The study also showed that many families are often shifting activities. There is also great variation in success in achieving commonly recognised desired livelihood outcomes.

In the study of the interrelationships between livelihood capital (natural, social, human, physical and financial), strategies and livelihood outcomes, only financial capital could be directly related to the measured livelihood outcomes. But the variance accounted for is small, 26%. Analysing the six components of the livelihood outcome separately reveals some associations. Those who have introduced animals score higher in physical capital, but lower in social capital. Those who have established businesses score low in social capital, i.e., farmers without a business depend more on groups. Number of children in secondary school is positively correlated with natural capital as older families have had more time to invest in natural capital. There is a marginal negative association between financial capital and farmers who have introduced crops. Farmers with off-farm income are not either using more fertiliser. These may show that off-farm income is not being used to invest in agriculture. Pulling the capitals apart revealed a strong association between off-farm jobs and the outcome. Number of animals also contributes clearly to the livelihood outcome. Older farmers have achieved more livelihood outcomes up to age 40, but the trend is not detectable thereafter. Further, access to roads is positively, and access to land in two ecological zones marginally positively associated with higher outcome scores. Despite the variation, financial capital may be the most crucial asset in influencing the realization of livelihood outcomes. The increasing need of cash is forcing farmers to look for off-farm jobs.

The SL framework works well as an organising tool for quantitative livelihood data collection. However, much in the livelihoods strategies and outcomes remains unexplained by a quantitative analysis using the SL livelihood framework. Insights into the unexplained can be sought from the many farm specific and changing combinations of capital and strategies; from the selection of indicators used in the analysis; and from the explanations and concepts offered by other livelihood models. Many intangible factors such as human capabilities, social relations and dynamics, preferences, perceptions and information capital can play a considerable role in explaining livelihood dynamics. They can be added to the framework. However, their inclusion in a quantitative analysis remains a challenge.

REFERENCES

Ashley, C. (2000). Applying livelihood approaches to natural resource management initiatives: Experiences in Namibia and Kenya. *Working Paper 134*. London: Overseas Development Institute.

Barahona, C., and Levy, S. (2003). How to generate statistics and influence policy using participatory methods in research: reflections on work in Malawi 1999-2002. *IDS Working Paper 212*. Brighton: Institute of Development Studies.

Barnum, H.N., and Squire, L. (1979). An econometric application of the theory of the farm household, *Journal of Development Economics, 6,* 79–102.

Barrett, C.B., Marenya, P.P., McPeak, J., Minten, B., Murithi, F., Oluoch-Kosura, W., Place, F., Randrianarisoa, J.C., Rasambainarivo, J., and Wangila, J. (2006). Welfare dynamics in rural Kenya and Madagascar. *Journal of Development Studies, 42,* 248–277.

Bebbington, A. (1999). Capitals and capabilities: A framework for analysing peasant viability, rural livelihoods and poverty. *World Development, 27,* 2021–44.

Becker, G.S. (1965). A theory of the allocation of Time. *Economic Journal, 75*, 493–517.

Bryceson, D.F. (1999). African rural labour, income diversification and livelihood approaches: A long term development perspective. *Review of African Political Economy, 80*, 171-189.

Carswell, G. (1997). Agricultural intensification and rural sustainable livelihoods: A 'think piece'. *IDS Working paper 64*. Brighton: Institute of Development Studies, University of Sussex.

CBS (2004). *Where are the poor? Geographic dimensions of well-being in Kenya* [CD-ROM]. Central Bureau of Statistics, Ministry of Planning and National Development in collaboration with The Rockefeller Foundation, ILRI (International Livestock Institute), The World Bank Group and World Resources Institute.

Carney, D. (1998). Implementing the sustainable rural livelihoods approach. In D. Carney (Ed.), *Sustainable Rural Livelihoods: What Contribution Can We Make?* (pp. 3-26). London: Department for International Development (DFID).

Chambers, R. (1987). Sustainable livelihoods, environment and development: Putting poor rural people first. *IDS Discussion Paper 240*. Brighton: Institute of Development Studies, University of Sussex.

Chambers, R., and Conway, G. (1992). Sustainable rural livelihoods: Practical concepts for the 21st Century. *IDS Discussion paper 296*. Brighton: Institute of Development Studies, University of Sussex.

Davies, S. (1996). *Adaptable livelihoods: Coping with food insecurity in the Malian Sahel.* London: Macmillan.

Deere, C.D., and de Janvry, A. (1979). A conceptual framework for the empirical analysis of peasants. *American Journal of Agricultural Economics, 61,* 601–11.

DFID (Department for International Development) (2001). *Sustainable livelihoods guidance sheets, Section 2.* London: Department for International Development.

Devereux, S. (1993). Goats before ploughs: Dilemmas of household response sequencing during food shortages. *IDS Bulletin 24*. Brighton: Institute of Development Studies, University of Sussex.

Eriksen, S.H., Brown, K., and Kelly, P.M. (2005). The dynamics of vulnerability: locating coping strategies in Kenya and Tanzania. *The Geographical Journal, 171,* 287-305.

Fleuret, A. (1988). Some consequences of tenure and agrarian reform in Taita, Kenya. In R.E. Downs and S.P. Reyna (Eds.), *Land and society in contemporary Africa* (pp. 136-158). Hanover and London: University Press of New England.

Haggblade, S., Hazell, S.P., and Brown, J. (1989). Farm-nonfarm linkages in rural sub-Saharan. *Africa World Development, 17,* 1173-1201.

Hamilton-Peach, J., and Townsley, P. (2004). *An IFAD sustainable livelihoods framework.* www.ifad.org/sla/framework/

Hobley, M. (2001). *Unpacking the PIP box.* Somerset, UK: Hobley Shields Associates. http://www.livelihoods.org/pip/pip/PIP-Syn.rtf.

Jaetzold, R., and Schmidt, H. (1983). *Farm management handbook of Kenya, 1983 Vol II/C East Kenya, Natural conditions and farm management information.* Ministry of Agriculture, Kenya, in cooperation with the German Agricultural Team (GAT) of the German Agency for Technical Cooperation (GTZ) Rossdorf, Germany.

Kenya, Republic of (1976). *Taita Taveta district development plan 1974-78.* Nairobi: Government Printer.

Kenya, Republic of (1989). *Taita-Taveta district development plan 1989-1993.* Nairobi: Government Printer.

Kenya, Republic of (1997). *Taita-Taveta district development plan 1997-2001.* Nairobi: Office of the Vice-President and Ministry of Planning and National Development Government Printer.

Livingstone, I. (1991). A reassessment of Kenya's rural and urban informal sector. *World Development, 19,* 651-670.

Mkangi, G. C. (1978). *Population growth and the myth of land reform in Taita.* PhD, Thesis. Brighton: University of Sussex.

Moser, C.O.N. (1998). The asset vulnerability framework: Reassessing urban poverty reduction strategies. *World Development, 26,* 1–19.

Nazzaro, A. A. (1974). *Changing use of the resource base among the Taita of Kenya.* PhD thesis. Michigan: Department of Geography, Michigan State University.

Odero, K. K. (2003). Extending the sustainable livelihoods framework. *Working Paper, Department of Rural and Urban Planning.* University of Zimbabwe. http://dlc.dlib.indiana.edu/archive/00001293/

ORNL (Oak Ridge National Laboratory), 2000. Global Ambient Population, 2000. http://sedac.ciesin.columbia.edu/plue/gpw/landscan/

Pattanayak, S., Mercer, S.E., Sills, E., and Yang, J. (2003). Taking stock of agroforestry adoption studies. *Agrforestry Systems, 57,* 173-186.

Reardon, T. (1997). Using evidence to household income diversification to inform study of the rural nonfarm labour market in Africa. *World Development, 25,* 735-747.

Scoones, I. (1998). Sustainable rural livelihoods: A Framework for analysis. *IDS Working Paper 72.* Brighton: Institute of Development Studies, University of Sussex.

Sen, A. (1981). *Poverty and famine.* Oxford: Oxford University Press.

Sen, A.K. (1985). *Commodities and capabilities.* Amsterdam: North-Holland.

da Silva, J. (2006). Annotated bibliography of recent Q2 analyses of poverty. *Q-Squared Working Paper No. 30.* Toronto: Centre for International Studies, University of Toronto.

Soini, E. (2005a). Land use change patterns and livelihood dynamics on the slopes of Mt. Kilimanjaro, Tanzania. *Agricultural Systems,* 85, 306-323.

Soini, E. (2005b). Livelihood capital, strategies and outcomes in the Taita Hills of Kenya. *ICRAF Working Paper no 8*. Nairobi: World Agroforestry Centre.

Soini, E. (2006). *Livelihood, land use and environment interactions in the highlands of East Africa,* Academic dissertation. Helsinki: Department of Geography, Faculty of Science, University of Helsinki.

Start, D., and Johnson, C. (2004). Livelihood options? The political economy of access: Opportunity and Diversification. *Working Paper 233.* London: Overseas Development Institute.

Swift, J. (1993). Understanding and preventing famine and famine mortality. *IDS Bulletin 24*. Brighton: Institute of Development Studies.

Swynnerton, R.J.M. (1954). *A plan to intensify the development of African agriculture in Kenya.* Nairobi: Government Printer.

In: Agricultural Systems: Economics, Technology and Diversity ISBN 978-1-60692-025-1
Editor: Oliver W. Castalonge

Chapter 7

BIO-PHYSICAL AND SOCIO-ECONOMIC DETERMINANTS OF ADOPTING A DIVERSIFIED CROPPING SYSTEM: A CASE STUDY FROM BANGLADESH

Sanzidur Rahman*
School of Geography, University of Plymouth, UK

ABSTRACT

The chapter explores the bio-physical and socio-economic determinants of adopting a diversified cropping system using a survey of 406 farmers located in 21 villages in three agro-ecological regions of Bangladesh. The computed value of the Herfindahl index of crop diversification confirms that farming system in Bangladesh is still relatively diverse despite four decades of thrust in the diffusion of "Green Revolution" technology package aimed at promoting modern rice monoculture. Although the gross value of output per hectare is significantly higher for specialized farms (i.e., farms concentrating on modern rice monoculture), the profits between specialized and diversified farms are similar because of significantly lower use rates of all variable inputs by the latter. Results from the Tobit model reveal that a host of factors, including input prices, significantly determines decision to adopt a diversified cropping system. Among the input prices, a rise in labour wage significantly promotes crop diversification, whereas a rise in the prices of fertilizers, pesticides and animal power services favour specialization towards modern rice monoculture. Crop diversification is positively associated with farm size. However, tenancy and availability of irrigation significantly favour specialization, implying that owner operators and farmers with no irrigation facilities are more likely to choose crop diversification. Both education and farming experience are significantly positively associated with crop diversification. Adoption of a diversified cropping system is significantly higher in regions endowed with developed infrastructure and relatively better soil fertility. Share of non-agricultural income also positively influence crop

* Address for correspondence: Dr. Sanzidur Rahman, Senior Lecturer in Rural Development, School of Geography, University of Plymouth, Drake Circus, Plymouth, PL4 8AA, Phone: +44-1752-585911, Fax: +44-1752-585998; E-mail: srahman@plymouth.ac.uk

diversification. Therefore, adoption of a diversified cropping system could be promoted significantly by investing in rural infrastructure, farmer education, and soil conservation measures. Price policies aimed at improving labour wage and reducing fertilizers, pesticides, and animal power prices would also promote crop diversification. In addition, land reform policies focusing on delegating land ownership to landless and marginal farmers are noteworthy.

Key words: diversified cropping system, crop diversification vs. specialization, biophysical factors, Tobit model, Bangladesh

1. INTRODUCTION

The economy of Bangladesh is largely dependent on agriculture. Although, rice production dominates the farming system of Bangladesh, accounting for 70% of the gross cropped area (BBS, 2001), several other crops are also grown in conjunction with rice in order to fulfil a dual role of meeting subsistence as well as cash needs. Since the beginning of 1960s, Bangladesh has pursued a policy of rapid technological progress in agriculture leading to diffusion of a rice-based 'Green-Revolution' technology package. As a result farmers concentrated on producing modern varieties of rice all year round covering three production seasons (*Aus* - pre-monsoon, *Aman* - monsoon and *Boro* - dry winter), particularly in areas that are endowed with supplemental irrigation facilities. This raised concern regarding the loss of crop diversity, consequently leading to an unsustainable agricultural system. For example, Husain et al., (2001) noted that the intensive monoculture of rice led to a displacement of land under low productive non-rice crops such as pulses, oilseeds, spices and vegetables, leading to an erosion of crop diversity, thereby, endangering the sustainability of crop-based agricultural production system. Mahmud et al., (1994: 02) also noted that "the area under non-cereal crops has continuously fallen since late 1970s, mainly due to the expansion of irrigation facilities, which led to fierce competition for land between modern *Boro* season (dry winter) rice and non-cereals". However, an analysis of the level of crop diversification between the two Agricultural Censuses of 1960 and 1996 reveals that the level of crop diversity has actually increased by 4.5 percent over the 36 year period (Table 1). Table 1 presents the cropped area, cropping intensity and the level of crop diversification between the Agricultural Censuses of 1960 and 1996, respectively. The former census just precedes the onset of Green Revolution while the latter census is the latest available to date, and is comprehensive in nature, scope and content. It is clear from Table 1 that, although there were dramatic changes in the structure of the farms and rise in cropping intensity, the level of crop diversification changed only moderately over a 36-year period.

With a boom in population growth between the two censuses, the average operational size per farm shrunk dramatically. The large and medium farms gave way to small farms largely because of the increase in the number of farms competing for a fixed amount of cultivable land. The decline in net cropped area by 12.7 percent is largely offset by a 26 percent rise in cropping intensity, thereby, leaving the gross cropped area slightly positive (a 2.5 percent increase). An examination of the crop shares reveals that the share of cereals (rice, wheat and other minor cereals) increased only by 1.7 percent in 36 years.

Table 1. Changes in cropped area, cropping intensity and diversification (1960 and 1996)

Indicators	Census 1960	Census 1996	Inter-census change (%)
Number of farms	6,139,480	11,798,242	92.17
% of small farms (0.02 – 1.01 ha)	51.63	79.87	197.26
% of medium farms (1.01 – 3.03 ha)	37.68	17.61	-10.19
% of large farms (above 3.03 ha)	10.69	2.52	-54.63
Operated area (ha)	7,744,929	8,076,369	4.28
Net temporary cropped area (ha)	7,627,372	6,655,771	-12.74
Gross cropped area (ha)	11,283,169	11,580,666	2.64
Operated area per farm (ha)	1.26	0.68	-45.74
Net temporary cropped area per farm (ha)	1.24	0.56	-54.59
Gross cropped area per farm (ha)	1.84	0.98	-46.59
Proportion of cropped area under (%)			
Rice	75.76	72.80	-1.37
Wheat and other minor cereals	0.83	5.52	585.35
Pulses	6.31	4.63	-24.72
Oilseeds	2.82	4.14	50.92
Cash crops	8.74	6.32	-25.77
Vegetables	2.22	3.55	64.49
Spices and other miscellaneous crops	3.33	3.04	-6.34
Cropping intensity (all farms)	148	174	26.00
Small farms	167	187	20.00
Medium farms	152	171	19.00
Large farms	135	154	19.00
Herfindahl index of crop diversification (all farms)	0.59	0.54	-4.50
Small farms	0.57	0.52	-4.59
Medium farms	0.59	0.55	-4.23
Large farms	0.60	0.59	-0.79

Source: Computed from BBS, 1999 and MoFA, 1962.

The main changes were in the composition of modern varieties of rice, which replaced the traditional varieties. Also, there was a five-fold increase in modern wheat area. The non-cereal crops used to account for 23.4 percent of the gross cropped area in 1960 and now accounts for 21.7 percent in 1996, a negligible decline of 1.4 percent in 36 years. However, there has been a shift in the composition of non-cereals over this period. The area under pulses, cash crops and spices declined sharply, while area under oilseeds and vegetables increased over this period. As a result, crop diversity in agriculture (as revealed by the computed Herfindahl index of crop diversification) increased by 4.5 percent over a 36 year period, which can hardly be justified as a serious replacement of land area by rice monoculture. The Herfindahl index of crop diversification was computed at 0.59 in 1960 and 0.54 in 1996.

There is an apparent paradox in that many non-cereal crops (e.g., potatoes, vegetables, onions and cotton) are more profitable (both in economic and financial terms) than modern rice cultivation, which was mainly attributed to high risk as well as incompatibility of the existing irrigation system to produce non-cereals in conjunction with rice (Mahmud et al., 1994). However, it has been increasingly recognized that, under non-irrigated or semi-irrigated conditions, better farming practices and varietal improvements in non-cereal crops will be more profitable and could lead to crop diversification as a successful strategy for the future growth and sustainability of Bangladeshi agriculture (MoA, 1989; Mahmud et al., 1994; PC, 1998). The Fifth Five Year Plan (1997–2002) set specific objectives to attain self-sufficiency in foodgrain production along with increased production of other nutritional crops, as well as to encourage the export of vegetables and fruits keeping in view domestic production and need (PC, 1998). The Plan also earmarked Tk 1,900 million (US$ 41.8 million) accounting for 8.9 percent of the total agricultural allocation to promote crop diversification. Such an emphasis at the policy level points towards the importance of identifying the determinants of farmers' crop choice decisions, so that an informed judgment can be made about the suitability of crop diversification as a desired strategy for promoting agricultural growth in Bangladesh.

Given this backdrop, the present chapter is aimed at determining the underlying biophysical and socio-economic factors affecting decision to adopt a diversified cropping system by Bangladeshi farmers. We use a Tobit model for this purpose, which has the advantage of taking into account all of the observations in the sample (McDonald and Moffit, 1980).

The rest of the chapter is divided into five sections. Section 2 provides the methodology including analytical framework and the sources of data. Section 3 presents the empirical model. Section 4 discusses the results. Section 5 provides conclusions and draws policy implications.

2. Methodology

2.1. Data and the Study Area

The study is based on farm-level cross section data for the crop year 1996 collected from three agro-ecological regions of Bangladesh. The survey was conducted from February to April 1997. Samples were collected from eight villages of the Jamalpur Sadar sub-district of Jamalpur, representing wet agro-ecology, six villages of the Manirampur sub-district of Jessore, representing dry agro-ecology, and seven villages of the Matlab sub-district of Chandpur, representing wet agro-ecology in an agriculturally advanced area. A multistage random sampling technique was employed to locate the districts, then the *Thana* (sub-districts), then the villages in each of the three sub-districts, and finally the sample households. A total of 406 households[1] from these 21 villages were selected. Detailed crop input-output data at the plot level for individual farm households were collected for ten crop groups[2]. The dataset also includes information on the level of infrastructural development[3]

[1] The sample households were selected based on the information on the total number of households including their land ownership categories, which were obtained from BRAC (a national non-governmental organization). Then a stratified random sampling procedure was applied using a formula from Arkin and Colton (1963) that maximizes the sample size with a 5% error limit. Farm size categories (large, medium, and small farmers) were used as the strata (for details, see Rahman, 1998).

[2] The crop groups are: traditional rice varieties (Aus – pre-monsoon, Aman – monsoon, and Boro – dry seasons), modern/high yielding rice varieties (Aus, Aman, and Boro seasons), modern/high yielding wheat varieties, jute, potato, pulses, spices, oilseeds, vegetables, and cotton. Pulses in turn include lentil, mungbean, and gram.

and soil fertility determined from soil samples collected from representative locations in the study villages[4].

2.2. The Theoretical Framework: The Tobit Model

Economic analysis of the farmers' technology adoption decision is deeply rooted on the assumption of utility maximization (e.g., Adesina and Baidu-Forson, 1995; and Adesina and Zinnah, 1993). The underlying utility function, which ranks the preference of individual farmers of a given technology, is not observable. What is observed is a set of farm and farmer specific socio-economic characteristics that influence farmers' decision to adopt a given technology, which is assumed to provide him/her with a certain level of perceived utility. Following this technology adoption paradigm, we postulate that, at the post adoption stage, an observable set of technology attributes and farm specific socio-economic characteristics will influence farmers' decision to adopt a diversified cropping system. This is because a farmer's adoption decision may be determined by his/her experience of growing various crops, extension visits, his/her knowledge about the technologies and other conditions (Negatu and Parikh, 1999).

2.3. The Econometric Model

Among the limited dependent variable models widely used to analyse farmers' decision making processes, Tobit analysis has gained importance since it uses all observations, both those are at the limit, usually zero (e.g., non-adopters), and those above the limit (e.g., adopters), to estimate a regression line, as opposed to other techniques that uses observations, which are only above the limit value (McDonald and Moffit, 1980). In our case, farmers could be highly diversified or concentrate only on specialization of modern rice monoculture. In such a case, the application of Tobit analysis is most suited because of the censored nature of the data. The stochastic model underlying Tobit may be expressed as follows (McDonald and Moffit, 1980):

Spices include onion, garlic, chilli, ginger, and turmeric. Oilseeds include sesame, mustard, and groundnut. Vegetables include eggplant, cauliflower, cabbage, arum, beans, gourds, radish, and leafy vegetables.

[3] A composite 'index of underdevelopment of infrastructure' was constructed using the cost of access approach. A total of 13 elements are considered for its construction. These are primary market, secondary market, storage facility, rice mill, paved road, bus stop, bank, union office, agricultural extension office, high school, college, thana (sub-district) headquarters, and post office (see Ahmed and Hossain, 1990 for construction details).

[4] The 'soil fertility index' was constructed from test results of soil samples collected from the study villages during the field survey. Ten soil fertility parameters were tested. These are soil pH, available nitrogen, available potassium, available phosphorus, available sulphur, available zinc, soil texture, soil organic matter content, cation exchange capacity of soil, and electrical conductivity of soil (for details of sampling and tests, see Rahman and Parkinson, 2007 and Rahman, 1998).

$$\begin{aligned} y_i &= X_i\beta + u_i && \text{if } X_i\beta + u_i > 0 \\ &= 0 && \text{if } X_i\beta + u_i \le 0, \\ & && i = 1,2,......n, \end{aligned} \tag{1}$$

where n is the number of observations, y_i is the dependent variable (Herfindahl index), X_i is a vector of independent variables representing technology attributes and farm and farmer specific socio-economic characteristics, β is a vector of parameters to be estimated, and u_i is an independently distributed error term assumed to be normal with zero mean and constant variance σ^2. The model assumes that there is an underlying stochastic index equal to $(X_i\beta + u_i)$ which is observed when it is positive, and hence qualifies as an unobserved latent variable. The relationship between the expected value of all observations, E_y and the expected conditional value above the limit E_y* is given by:

$$E_y = F(z)\, E_y*$$

where F(z) is the cumulative density normal distribution function and $z = X\beta/\sigma$. The full maximum likelihood estimation procedure is utilized using NLOGIT-4 (ESI, 2007) software program.

3. The Empirical Model

A Tobit model is developed to empirically investigate the socio-economic factors underlying the decision to adopt a diversified cropping system. Two classes of variables can be used to represent the specialization variable. The Herfindahl index, which is based on the area allocated to a particular enterprise (Llewelyn and Williams, 1996) or the Ogive index, which is defined as a concentration of output shares of various enterprises (Coelli and Fleming, 2004). The Herfindahl index is represented as: $DV = \sum_{i=1}^{n} P_i^2$, where P_i is the proportion of farm area involved in a particular enterprise. The value of Herfindahl index ranges between 0 and 1 with 0 denotes perfect diversification and 1 denotes perfect specialization. In this study, we have selected the Herfindahl index of crop diversification[5]. The justification is that land is the scarcest input in Bangladesh compared to any other resource requirements. The land-person ratio is one of the lowest in the world, estimated at only 0.12 ha (FAO, 2001).

Several studies have analyzed the determinants of modern technology adoption by farmers in developing countries using OLS, probit or Tobit regressions of technology adoption on the variables representing: (a) socio-economic circumstances of farmers – such as, farm size, tenurial status, farmers' education level, farming experience, family size, gender, etc., and (b) institutional and physical infrastructures – such as, irrigation, credit,

[5] We have also analysed the data using the Ogive index of output concentration, which provided almost identical results. The correlation coefficient between the two indices is estimated at 0.98 ($p<0.01$).

extension contact, membership in organizations, distance to market/bus stop/extension office, etc.. (e.g., Hossain, 1989; Nkamleu and Adesina, 2000; Shiyani, et al., 2002; Floyd et al., 2003; and Ransom, et al., 2003). However, most of these studies ignored or omitted price factors as determinants of technology adoption, which has important bearing on profitability and/or resource allocation decisions and, hence, provides an incomplete picture of farmers' decision-making process. Therefore, we have added input prices in addition to variables representing bio-physical and socio-economic circumstances of the farmers to examine their individual influences on decision to diversify.

The variables incorporated in the Tobit model were: four variable input prices (i.e., fertilizers, pesticides, labour and animal power services), amount of land cultivated, value of farm capital assets, irrigation access, tenurial status, farmers' education, farming experience, subsistence pressure, extension contact in the past one year, share of non-agricultural income, index of underdevelopment of infrastructure, soil fertility index, and the environmental awareness index[6]. The definition and measurement of all these variables are presented in Table 4. The justification for including these variables in the model is discussed below.

Land is the scarcest resource in Bangladesh, and the farm size largely determines the level and extent of income to be derived from farming. Land also serves as a surrogate for a large number of factors as it is a major source of wealth and influences decision to choose crops. However, the impact of tenancy on the extent of modern rice technology adoption among farmers is varied (Hossain, et al., 1990). Hence, the amount of land cultivated (to represent wealth) and the proportion of land rented-in (to represent tenurial status) were incorporated to test their independent influence on decisions regarding crop choices.

Farmers in Bangladesh are not only land poor, but also resource poor. The farm capital asset variable (which includes the value of all tools and equipments as well as the livestock resources used directly in the production process) was included in order to examine whether farm capital has a bearing on crop choice decisions.

Access to modern irrigation facilities is an important pre-requisite for growing modern rice varieties, particularly the HYV *Boro* rice grown in dry winter season. Lack of access to modern irrigation facilities has been identified as one of the principal reasons for stagnation in the expansion of modern rice area, which currently accounts for a little over 50 percent of total rice area (Rahman and Thapa, 1999; Mahmud et al., 1994).

Use of farmers' education level as explanatory variable in adoption studies is common (e.g., Nkamleu and Adesina, 2000; Adesina and Baidu-Forson, 1995). The education variable was used as a surrogate for a number of factors. At the technical level, access to information as well as capacity to understand the technical aspects and profitability related to different crops may influence crop production decisions. The justification of including farming experience is straightforward. Experienced farmers are more likely to be open to choices regarding crops, be it modern rice or non-rice crops.

Agricultural extension can be singled out as one of the important sources of information dissemination directly relevant to agricultural production practices, particularly in nations like Bangladesh where farmers have very limited access to information. This was reinforced by

[6] An 'environmental awareness index' was constructed based on the farmers' perception of the impacts of modern agricultural technology diffusion in agriculture (i.e., modern rice production). A set of 12 specific environmental impacts were read to the respondents who were asked to reveal their opinion on each of these impacts as well as to rank their importance on a scale of 1 to 5 (for details of the construction procedure, see Rahman, 2005; 2003).

the fact that many studies found a significant influence of extension education on adoption of land-improving technologies (e.g., Adesina and Zinnah, 1993). Therefore, this variable was incorporated to account for its influence on adoption decisions.

According to Chayanovian theory of the peasant economy, higher subsistence pressure increases the tendency to adopt new technology, and this has been found to be the case in Bangladesh (Hossain, et al., 1990; Hossain, 1989). The subsistence pressure variable, measured by family size per household was incorporated to account for its influence on crop choices.

The percentage of income earned off-farm was included to reflect the relative importance of non-agricultural work in these farm households. It may also reflect farmers' increased ability to meet operational cost, particularly when choosing high value cash crops.

Infrastructure affects agricultural production indirectly through prices, diffusion of technology and use of inputs and has profound impact on the incomes of the poor (Ahmed and Hossain, 1990). The state of infrastructure implies improved access to markets and institutions as well as better access to information and hence may influence farmers' crop choices. This effect was captured by the index of underdevelopment of infrastructure.

Soil fertility is a key factor that exerts a positive influence on productivity (e.g., Rahman, 2005; Rahman and Parkinson, 2007), which in turn may influence decision to choose crops.

Rahman (2003, 2005) noted that Bangladeshi farmers' are aware of the adverse environmental impacts caused by modern technology diffusion, and such awareness has subsequent impact on their resource allocation decisions. Hence, the 'environmental awareness index' variable was included in this study to examine whether it has an influence in decision to choose a diversified cropping system.

4. Results

4.1. Level of Crop Diversification

Table 2 presents the existing cropping practice and the extent of crop diversification amongst the sampled households in each region.

It is clear from Table 2 that there are substantial variations among the regions with respect to each of the aspects considered. Although 51 percent of the total farmers adopted modern rice monoculture, a substantial 37 percent of the total farmers adopted both modern rice as well as a diversified cropping system. In terms of area allocated to crops, the non-rice crops cover an estimated 19 percent of gross cropped area.

In fact, farmers produce a wide range of crops in a cropping year. The mean number of crops grown is estimated at 3.6 with a maximum of 11 crops in a year. The lower panel of Table 2 presents the actual measure of crop diversification using the Herfindahl index.

The overall Herfindahl index of crop diversification is estimated at 0.60, which indicates that cropping system in Bangladesh is relatively diverse, particularly in Jessore region, where the level of modern rice technology adoption is lowest.

Table 2. Extent of crop diversification among sampled farmers

Variables	Comilla	Jessore	Jamalpur	All region
Proportion of farmers:				
Only modern rice adopter	0.51	0.27	0.65	0.51
Only diverse crop adopter	0.16	0.23	0.02	0.12
Adopter of both diversified crop and modern rice	0.33	0.50	0.33	0.37
Proportion of gross cropped area under:				
Modern rice only	0.65	0.32	0.63	0.56
Diverse crops (excluding all types of rice)	0.22	0.37	0.07	0.19
Traditional rice only	0.13	0.31	0.30	0.25
Average number of crops grown in one year	3.34	4.19	3.35	3.57
Standard deviation	1.57	2.16	1.73	1.85
Maximum number of crops grown in one year	8.00	11.00	10.00	11.00
Crop diversification index				
Herfindahl index of crop diversification	0.69	0.46	0.63	0.60
Number of observations (farm households)	126	105	175	406

Note: The actual data were collected at plot level. Therefore, the total plot level observations of all types of crops grown by these 406 farmer stands at 1,448. Number of observations of modern rice = 622 (Aus = 25, Aman = 150, and Boro = 447); traditional rice = 324 (Aus = 37, Aman = 266, and Boro = 21); and diverse crops = 502 (wheat = 103, jute = 92, potatoes = 59, pulses = 70, spices = 47, oilseeds = 71, vegetables = 44, and cotton = 16). Pulses in turn include lentil, mungbean, and gram. Spices include onion, garlic, chilly, ginger, and turmeric. Oilseeds include sesame, mustard, and groundnut. Vegetables include eggplant, cauliflower, cabbage, arum, beans, gourds, radish, and leafy vegetables.

Higher index values of crop diversification indicate specialization (i.e., in this case towards modern rice monoculture).

Table 3. Profitability and input use rates of diversified and specialized farms

Variables	Diversified farms (Herfindahl index <0.60)	Specialized Farms (Herfindahl index > 0.9)	Mean difference (Diversified vs. Specialized)	t-ratio
Land area cultivated (ha)	1.10	0.62	0.48	4.21***
Labour (days/ha)	91.11	123.30	-32.19	-7.01***
Animal power services (days/ha)	26.07	30.32	-4.24	-4.64***
Fertilizer (kg/ha)	210.06	262.33	-52.27	-6.46***
Pesticides (Taka/ha)	298.83	672.94	-374.12	-6.37***
Irrigation (Taka/ha)	1585.59	1522.42	63.17	0.47
Seed (kg/ha)	116.49	108.12	8.37	0.79
Gross value of output (Taka/ha)	22,213.31	24,416.29	-2,202.98	-2.76***
Profits (Taka/ha)	13,065.26	13,606.92	-541.66	-0.73
Number of farms	260	108		

Note: Profits = (gross value of output – variable cost of all inputs).

*** = significant at 1 percent level ($p<0.01$).

Table 4. Summary statistics of the variables used in the Tobit model

Variables	Unit of measurement	Mean	Standard deviation
Price variables			
Animal power price	Taka per animal pair-days	83.56	17.34
Fertilizer price	Taka per kg	5.85	1.38
Labour wage	Taka per person-day	44.67	8.22
Pesticide price	Taka per hectare	1015.14	1300.95
Biophysical and socio-economic factors			
Amount of land cultivated	Hectare	0.98	1.02
Farm capital asset	Thousand taka	55.38	116.85
Irrigation	Proportion of land under irrigation	0.62	0.30
Tenurial status	Proportion of land rented in	0.20	0.29
Education of farmer	Completed year of schooling	3.74	4.26
Farming experience	Years	25.51	14.21
Family size	Persons per household	6.02	2.53
Infrastructure index	Number	33.32	14.95
Soil fertility index	Number	1.68	0.19
Environmental awareness index	Number	0.37	0.17
Extension contact	Dummy (1 if had contact, 0 otherwise)	0.13	0.33
Share of non-agricultural income	Proportion of total household income	0.22	0.31
Herfindahl index	Number	0.60	0.27
Number of observations		406	

4.2. Profitability and Input Use Rates of Diversified Farms

Table 3 presents input use rates classified by the level of farm diversification. We designated farms below the mean level of the Herfindahl index of the sample (<0.60) as 'diversified farms' and farms with the Herfindahl index>0.90 as 'specialized farms', who were largely modern rice producers. It is clear from Table 3 that the operational size of diversified farms is significantly higher and the use rates of inputs per hectare, except irrigation and seeds, are significantly lower. The use rates of labour, fertilizer and pesticide are 35, 24 and 125 percent lower among diversified farms compared with those of specialized farms. Although, gross value of output is significantly higher for specialized farms, the profits are similar between specialized and diversified farms, due to significantly lower use of inputs by the latter.

4.3. Determinants of Adopting a Diversified Cropping System

Summary statistics of the variables used in the Tobit model are presented in Table 4. The farm-specific variables provide a summary of the characteristics of these farms. The amount of land cultivated per farm is 0.98 ha. The average level of education is less than four years; experience in farming is 26 years; average family size is six persons; 22 percent of income is derived off-farm; and only 13 percent of farmers have had contact with extension officers during the past year.

Table 5 presents the parameter estimates of the full information maximum likelihood estimation of the Tobit model.

Table 5. A Tobit analysis of the determinants of adopting a diversified cropping system (dependent variable: Herfindahl index)

Variables	Coefficient	t-ratio
Constant	0.4221	2.92***
Price variables		
Animal power price	0.0018	2.13**
Fertilizer price	0.0369	4.01***
Labour wage	-0.0053	-3.05***
Pesticide price	0.0002	2.24**
Biophysical and socio-economic factors		
Amount of land cultivated	-0.0243	-1.65*
Farm capital asset	-0.0001	-1.05
Irrigation	0.1814	4.42***
Tenurial status	0.0922	2.20**
Education of farmer	-0.0061	-1.98**
Farming experience	-0.0020	-2.29**
Family size	-0.0034	-0.70
Infrastructure index	0.0053	6.35***
Soil fertility index	-0.1001	-1.65*
Environmental awareness index	0.0810	1.08
Extension contact	-0.0396	-1.12
Share of non-agricultural income	-0.0602	-1.65*
Model diagnostics		
Log likelihood	43.69	
Number of observations	406	

Note: *** = significant at 1 percent level ($p<0.01$).
** = significant at 5 percent level ($p<0.05$).
* = significant at 10 percent level ($p<0.10$).

The coefficients on 12 out of a total of 16 variables were significantly different from zero at 10 percent level at least, indicating that the variables included in the model to explain the decision to adopt crop diversification (specialization) were correctly justified. Coefficients on all the four price variables were significantly different from zero, thereby, justifying the inclusion of price factors in studies examining technology adoption decisions, which was not commonly seen in the literature. A rise in labour wage significantly increases decision to adopt a diversified cropping system[7], whereas a rise in the remaining three input prices favour specialization towards modern rice monoculture. This is because farmers can spread the use of labour among different crops in a diversified cropping system, whereas in modern rice monoculture, large number of labourers is required within a very short period, particularly during sowing/transplanting and harvesting period, which were largely met by hired labour. Therefore, as labour wage rises, farmers opt for crop diversification.

Among the biophysical and socio-economic factors, farm size is positively related to the probability of choosing a diversified cropping system, as expected. The implication is that, as farm size increases, farmers are able to choose a diversified portfolio of crops which presumably meets the subsistence requirements (mainly rice) as well as generate surpluses for the market by growing high value non-cereal and/or cash crops. Availability of irrigation is the single most important determinant of specialization towards modern rice monoculture, as expected. The value of the coefficient on this variable is the highest in the model. This result corroborate with the finding of Hossain et al., (1990), who noted that access to irrigation is a major determinant of modern rice technology adoption in Bangladesh. In other words, it implies that the likelihood of adopting a diversified cropping system is higher in areas with no irrigation, which again corroborates with the conclusions of Mahmud et al., (1994) and Morris et al, (1996). In fact, wheat provides highest returns in non-irrigated zones and in areas that are unsuitable for *Boro* rice (Morris et al., 1996). Tenants are also more likely to choose specialization towards modern rice monoculture, which corroborates with the finding of Hossain et al., (1990). Also the tenurial system in Bangladesh is largely based on arrangements related to rice production. In the most common tenurial arrangement practiced in Bangladesh, the landlord receives one-third of the crop output share (mostly rice). The incidence of input cost share by the landlord varies across regions. Areas where such cost is shared (usually on a 50-50 basis), the arrangement is based on sharing of relatively scarce input, e.g., fertilizer, irrigation and/or animal power hire costs (Rahman, 1998). Therefore, existing tenurial arrangement seems to work well when the tenant grows rice. However, when a diversified cropping system is adopted, it may exert a discouraging effect, because the amount to be received as output share cannot be clearly estimated *a priori.*

Both the education level of farmer and the farming experience have a significant positive relationship with the decision to adopt a diversified cropping system, as expected. As mentioned earlier, the ability to process information increases with education as well as experience. Therefore, the educated and/or experienced farmers choose to adopt a diversified cropping system in order to take advantage of all the potentials arising from making such a choice, e.g., high returns for a particular crop, low overall resource cost, and/or spreading of scarce family labour evenly over a crop year.

The likelihood of adopting a diversified cropping system is significantly higher in regions with developed infrastructure[8]. The influence of developed infrastructure in adopting a diversified cropping system is obvious. For example, vegetable production provides a significantly higher return (Rahman, 1998), but is highly perishable and needs to be marketed immediately after harvest. The prospect of doing so increases only in regions with developed infrastructure. The likelihood of deciding in favour of crop diversification is higher in regions with fertile soils. This is because higher soil fertility implies higher productivity. And higher productivity of high value non-cereals, therefore, would probably be more profitable than modern rice monoculture.

The non-agricultural income share significantly influences probability to adopt a diversified cropping system. This is because higher non-agricultural income share reflects greater liquidity, thereby, supporting farmers to adopt a diversified cropping system which is characterized with a varied operational cost at different points of time during the production cycle.

[7] The Herfindahl index is a specialization index. Therefore, a negative coefficient on the explanatory variable implies positive relationship with diversification and vice-versa.

[8] The index reflects the underdevelopment of infrastructure, and therefore, a negative sign indicates positive effect on the dependent variable. In other words, the positive sign on the coefficient implies positive relationship towards crop diversification.

4.4. Discussions

Results of this study clearly reveal that a host of price and non-price factors influence the decision to adopt a diversified cropping system. These are labour wage, farmers' education and experience, farm operation size, regions endowed with developed infrastructure and better soil fertility and the share of non-agricultural income. When a farm diversifies into a combination of subsistence and cash crop production, the farmer uses the opportunity to select enterprises that complement each other given the nature of seasonality in demand for labour in particular. For instance, modern rice production exerts significant pressure on labour requirements during both the transplanting and harvesting seasons.

The cropping system in Bangladesh is largely influenced by access to water. The cropping pattern can be broadly classified into cropping under rainfed and irrigated conditions, which again vary according to the degree of seasonal flooding. As mentioned earlier, an apparent paradox exists in that, although many non-cereals are more profitable than producing modern rice, their expansion has stagnated due to the incompatibility of the existing modern irrigation systems (Mahmud et al., 1994). In fact, areas where modern irrigation is non-existent or unreliable, modern wheat is the desired crop and provides higher profitability (Morris et al., 1996). In general, the proportion of non-cereal crops is lower under irrigated conditions as compared to rainfed conditions (Mahmud et al., 1994). The sample households of this study also demonstrated that the cropping system is highly diverse in areas with poor irrigation facilities. For example, cropping diversity is significantly lower in the Comilla region in comparison to the Jamalpur and Jessore regions. This is because some of the villages in Comilla region fell within the Meghna-Dhonagoda Flood Control, Drainage and Irrigation (FCD/I) project, which resulted in the dominance of modern rice monoculture throughout the crop year because of the assured availability of water for irrigation at a cheap rate (Rahman, 1998).

An important issue that limits the scope to expand non-cereals is the existence of the price risk associated with uncertainties in marketing, particularly for perishable crops such as vegetables. In fact, annual variability in harvest prices is as high as 15–25 percent for most fruits and vegetables (including potatoes) and 20–40 percent for spices as compared to only 5–6 percent for cereals (Mahmud et al., 1994). This perhaps explains the decline in the area under spices between the census years (Table 1). Mahmud et al., (1994) further noted that the price shock is most severe at the level of primary markets during harvest seasons. Delgado (1995) stressed the need for addressing marketing issues and constraints as a priority option to promote agricultural diversification in sub-Saharan African regions. This is because in the absence of improved markets, the agricultural sector is likely to suffer from demand constraints as well as a weak supply response, thereby, affecting growth. One way to lower the price risk is through improvements in marketing, which in turn depends on the development of the rural infrastructure.

Result from this study clearly reveals that infrastructure significantly promotes adoption of a diversified cropping system. Infrastructure development in turn may also open up opportunities for marketing, storage and resource supplies, which would complement crop diversification. For example, Ahmed and Hossain (1990) concluded that farms in villages with relatively developed infrastructure use relatively greater amounts of fertilizer and market a higher percentage of their agricultural products in Bangladesh. Evenson (1986) noted a strong relationship between roads and increased agricultural production in the Philippines. He claimed that a 10 percent increase in roads would lead to a 3 percent increase in production in the Philippines. Ahmed and Donovan (1992: 31) concluded that "the degree of infrastructural development is in reality the critical factor determining the success of market-oriented sectoral and macroeconomic policies in the developing world".

Also, it should be noted that the non-cereals produced by most farmers comprised largely traditional varieties, which are low yielding. Strategy to improve varieties of non-cereals, therefore, provides further potential to improve productivity gains from diversification. Conventionally, the research and development activities in Bangladesh are largely concentrated on developing modern rice varieties to the neglect of most other crops. Among the non-cereals, modern technology is only well established in potato cultivation (Mahmud et al., 1994). The Bangladesh Agricultural Research Institute (BARI) is entrusted with the responsibility of developing modern varieties of all cereal and non-cereal crops except rice and jute. To date, a total of 131 improved varieties of various cereal and non-cereal crops have been developed and released by BARI (Appendix Table A1). A close look at the Appendix Table A1 reveals that only a third of these varieties were developed and released prior to the latest available Agricultural Census of 1996. The thrust in developing and releasing improved varieties by BARI seems to have gained momentum from mid-1990s, a complementary effort to government's emphasis on promoting crop diversification in its Fifth Five Year Plan document (1997 – 2002). However, there is a need to examine the impact of these new releases on farmers' portfolios of crop choices at the farm level, because the technical and socio-economic constraints on the diffusion of these technologies remain unexplored and less understood (Mahmud et al., 1994).

5. Conclusion

The aim of this study was to explore the biophysical and socio-economic determinants of adopting a diversified cropping system by farmers in Bangladesh using a Tobit model. Results reveal that a host of price and non-price factors influence the decision to adopt a diversified cropping system. Among the input prices, a rise in labour wages favour crop diversification. Availability of irrigation is the single most important determinant of specialization towards modern rice monoculture. Also tenants are likely to choose specialization in modern rice cultivation. On the other hand, decision to adopt a diversified cropping system is significantly positively related to education of the farmer, farming experience, farm operation size, regions endowed with developed infrastructure as well as better soil fertility, and the share of non-agricultural income.

5.1. Policy Implications

The key policy implications that emerges from this study is that crop diversification can be promoted significantly by investing in rural infrastructural development, soil conservation as well as education targeted for the farming population. Also, price policies aimed at improving labour wages and reducing fertilizers, pesticides, and animal power prices would positively influence crop diversification. A diversified cropping system is likely to have a positive impact on agricultural sustainability, as it is clear from the literature that the Green Revolution technology based on modern rice monoculture is unsustainable in the long-run.

Appendix Table A1. Improved varieties of crops released by BARI

Name of crop	Year of release	Yield (ton/ha)
VEGETABLES		
BARI Tomato-2	1985	85-90
BARI Tomato-3	1996	85-90
BARI Tomato-8	1998	100-115
BARI Tomato-9	1998	85-90
BARI Tomato-10 (S)	1998	45-50
BARI Tomato-13 (S)	1999	45-50
BARI Hybrid Tomato-3 (S)	2006	40-45
BARI Hybrid Tomato-4 (S)	2006	40-45
BARI Begun -1 (Brinjal)	1985	60.0
BARI Begun -3 (Brinjal)	1992	80.0
BARI Begun -4 (Brinjal)	1998	65.0
BARI Begun -5 (Brinjal)	1998	45-65
BARI Begun -6 (Brinjal)	2006	30-35
BARI Begun -7 (Brinjal)	2006	40-45
BARI Begun -8 (Brinjal)	2006	40-45
Tasaki mula-1 (Radish)	1996	75.0
BARI Mula-2 (Radish)	1996	60.0
BARI Mula-3 (Radish)	1998	55.0
BARI Phulkopi-1 (Cauliflower)	1998	28-30
BARI Phulkopi-2 (Cauliflower)	2006	25-27
BARI Lau-1 (Bottle gourd)	1996	42-45
BARI Lau-2 (Bottle gourd)	2006	55-60
BARI Seem-1 (Country bean)	1996	20-22
BARI Seem-2 (Country bean)	1996	11-12
BARI Seem-3 (Country bean)	2006	15.0
BARI Jhar Seem-1 (French bean)	1996	13-14
BARI Motor shuti-1 (Garden pea)	1996	10-12
BARI Motor shuti-2 (Garden pea)	1996	12-14
BARI Dherosh-1 (Okra)	1996	14-17
BARI Gimakalmi (Ipomoea reptans)	1983	40-45
BARI Puisak-1 (Indian Spinach)	1999	45-50
BARI Puisak-2 (Indian Spinach)	2006	55-60
BARI Lalsak-1 (Red amaranth)	1996	12-14
BARI Karola-1 (Bitter gourd)	2006	25-30
BARI Chalkumra-1 (Gourd)	2006	25-30
BARI Potol-1	2006	30.0
BARI Potol-2	2006	35-40
BARI Barbati-1 (Long bean)	2006	15-20
BARI Danta Sak-1	2006	18-20
BARI Danta Sak-2	2006	30-32

Appendix Table A1. (Continued)

Name of crop	Year of release	Yield (ton/ha)
WHEAT		
Protiva	1993	3.8-4.5
Kanchan	1983	3.5-4.6
BARI GAM-19 (Sourav)	1998	3.5-4.5
BARI GAM-20 (Gourab)	1998	3.6-4.8
BARI GAM-21 (Shatabdi)	2000	3.6-5.0
BARI GAM-22 (Sufi)	2005	3.6-4.8
BARI GAM-23 (Bijay)	2005	4.3-5.0
BARI GAM-24 (Prodip)	2005	3.5-5.1
OILSEEDS		
Tori 7 (Tori rape)	1976	0.9-1.1
Kalanyia (TS-72) (Tori rape)	1979	1.2-1.4
BARI Sarisa 9 (Tori rape)	2000	1.2-1.4
BARI Sarisa 12 (Tori rape)	2000	1.2-1.4
Sonali Sarisa (SS-75) (Tori rape)	1979	1.8-2.0
BARI Sarisa 6 (Tori rape)	1994	2.1-2.5
BARI Sarisa 7 (Rape seed)	1994	2.0-2.5
BARI Sarisa 8 (Rape seed)	1994	2.0-2.5
Rai-5 (Indian mustard)	1976	1.0-1.2
Daulat (Mustard)	1988	1.2-1.4
BARI Sarisa 10 (Mustard)	2000	1.2-1.4
BARI Sarisa 11 (Mustard)	2001	2.0-2.4
BARI Til 2 (Sesame)	2001	1.2-1.4
BARI Til 3 (Sesame)	2001	1.2-1.4
Maizchar Badam (Groundnut)	1976	1.6-2.0
Tridana Badam (Groundnut)	1987	2.0-2.2
BARI Chinabadam-5 (Groundnut)	1998	2.7-3.0
BARI Chinabadam-6 (Groundnut)	1998	1.8-2.0
Kironi (Sunflower)	1982	1.3-1.8
BARI Surjomukhi-2 (Sunflower)	2004	1.4-2.0
Shohag (Soybean)	1990	1.6-1.8
Bangladesh Soybean-4 (Soybean)	1994	1.6-2.5
MAIZE		
Barnali	1986	4.0-6.0
Khoibhutta	1986	2.5-4.0
Mohar	1990	3.5-5.5
BARI MAIZE-5	1997	3.5-6.5
BARI MAIZE-6	1998	5.0-7.0
BARI MAIZE-7	2002	5.0-7.5
BARI Sweet corn-1	2002	14.0
BARI Hybrid maize-1	2000	6.5-8.5
BARI Hybrid maize-2	2002	7.0-9.0
BARI Hybrid maize-3	2002	9.0-9.5
BARI Hybrid maize-4	2002	7.4-8.5
BARI Hybrid maize-5	2004	7.5-10
Shuvra	1986	5.5-6.0
PULSES		
BARI Masur-1 (Lentil)	1991	1.7-1.8
BARI Masur-2 (Lentil)	1993	1.5-1.7
BARI Masur-3 (Lentil)	1996	1.5-1.7

Appendix Table A1. (Continued)

Name of crop	Year of release	Yield (ton/ha)
BARI Masur-4 (Lentil)	1996	1.6-1.7
BARI Chola-1 (Chickpea)	1987	1.3-1.6
BARI Chola-2 (Chickpea)	1993	1.3-1.6
BARI Chola-3 (Chickpea)	1993	1.8-2.0
BARI Chola-4 (Chickpea)	1996	1.9-2.0
BARI Chola-5 (Chickpea)	1996	1.8-2.0
BARI Chola-6 (Chickpea)	1996	1.8-2.0
BARI Chola-7 (Chickpea)	1998	2.0-2.5
BARI Chola-8 (Chickpea)	1998	1.5-1.8
BARI Mung-2 (Mungbean)	1987	0.9-1.1
BARI Mung-3 (Mungbean)	1996	1.2-1.3
BARI Mung-4 (Mungbean)	1996	1.3-1.4
BARI Mung-5 (Mungbean)	1997	1.4-1.5
BARI Mung-6 (Mungbean)	2003	1.5-1.6
BARI Mash-1 (Blackgram)	1990	1.5-1.6
BARI Mash-2 (Blackgram)	1996	1.5-1.9
BARI Mash-3 (Blackgram)	1996	1.4-1.5
BARI Khesari-1 (Khesari)	1995	1.5-1.7
BARI Khesari-2 (Khesari)	1996	1.5-2.0
BARI Falon-1 (Falon)	1993	1.1- 1.4
SPICES		
BARI Onion-1	1996	12-16
BARI Onion-2	2000	10-13
BARI Onion-3	2000	10-13
BARI Garlic-1	2004	6-7
BARI Turmeric-1	1985	28-32
BARI Turmeric-3	2000	25-30
BARI Chilli-1	2001	2.5-3.0
BARI Coriander-1	1996	1.7-2.0
BARI Fenugreek-1	2000	1.2-1.5
TUBERS		
BARI Alu-1 (Potato)	1990	30-40
BARI Alu-7 (Potato)	1993	15-30
BARI Alu-8 (Potato)	1993	25-40
BARI Alu-12 (Potato)	1993	20-35
BARI Alu-13 (Potato)	1994	25-30
BARI Alu-17 (Potato)	2000	25-30
BARI TPS-1 (True Potato Seed)	1997	45-60
BARI SP-1 (Sweet Potato)	1985	40-45
BARI SP-2 (Sweet Potato)	1985	40-45
BARI SP-3 (Sweet Potato)	1988	30-35
BARI SP-4 (Sweet Potato)	1994	40-45
BARI SP-5 (Sweet Potato)	1994	35-40
BARI SP-6 (Sweet Potato)	2004	30-50
BARI SP-7 (Sweet Potato)	2004	30-50
Bilashi (Aroids)	1988	25-30
Latiraj (Aroids)	1988	25-30

Note: BARI also produce improved varieties of fruits and flowers which are not included in this list.
Source: Compiled from Hossain et al., (2006).

Therefore, the present thrust at the planning level to promote crop diversification is a step in the right direction. Another significant factor influencing crop diversification decision is the share of non-agricultural income of the household, which in turn improves with the development of rural infrastructure. Ahmed and Hossain (1990) concluded that infrastructure has profound impacts on the income of the poor in Bangladesh raising their income by 33 percent (which includes doubling of wages and increase in income from business and industries by 17 percent), thereby reinforcing our argument to improve rural infrastructure. Furthermore, appropriate land reform policies that focus on delegating land ownership to landless, marginal and small farmers as well as improvements in existing tenurial system, which is now biased towards favouring modern rice monoculture, would boost the number of owner operators, who are the most likely adopters of a diversified cropping system.

REFERENCES

Adesina, A.A., Baidu-Forson, J. 1995. Farmers' perception and adoption of new agricultural technology: evidence from analysis in Burkina Faso and Guinea, West Africa. *Agricultural Economics*. 13: 1–9.

Adesina. A.A., Zinnah, M.M., 1993. Technology characteristics, farmers' perceptions and adoption decisions: a Tobit model application in Sierra Leone. *Agricultural Economics*. 9: 297–311.

Ahmed, R., Donovan, C. 1992. Issues of infrastructural development: a synthesis of the literature. International Food Policy Research Institute, Washington, D.C.

Ahmed, R., Hossain, M., 1990. Developmental impact of rural infrastructure in Bangladesh. IFPRI Research Report 83. International Food Policy Research Institute, Washington, D.C.

Arkin, H., Colton, R.R. 1963. *Tables for Statisticians*. 2nd Edition. Harper and Row Publishers, New York.

BBS, 1999. *Census of agriculture – 1996. Structure of Agricultural Holdings and Livestock Population*. Volume 1. Bangladesh Bureau of Statistics, Dhaka.

BBS, 2001. *Statistical Yearbook of Bangladesh, 2000*. Bangladesh Bureau of Statistics, Dhaka.

Coelli, T.J., Fleming, E., 2004. Diversification economies and specialization efficiencies in a mixed food and coffee smallholder farming system in Papua New Guinea. *Agricultural Economics*. 31: 229-239.

Delgado, C.L. 1995. Agricultural diversification and export promotion in sub-Saharan Africa. *Food Policy*, 20: 225-243.

ESI, 2007. *NLOGIT-4*, Econometric Software, Inc. New York.

Evenson, R.E., 1986. Infrastructure, output supply and input demand in Philippine agriculture: provisional estimates. *Journal of Philippine Development*. 13: 62-76.

FAO, 2001. *Supplements to the Report on the 1990 World Census of Agriculture. Statistical Department Series (9a)*. Food and Agriculture Organization, Rome. http://www.fao.org/es/ess/census/wcahome.htm

Floyd, C., Harding, A-H., Paudel, K.C., Rasali, D.P. Subedi, K., Subedi, P.P., 2003. Household adoption and the associated impact of multiple agricultural technologies in the western hills of Nepal. *Agricultural Systems*, 76: 715–738.

Hossain, M., 1989. *Green Revolution in Bangladesh: Impact on Growth and Distribution of Income.* Dhaka: University Press Limited.

Hossain, M., Quasem, M.A., Akash, M.M., Jabber, M.A., 1990. Differential impact of modern rice technology: the Bangladesh case. Bangladesh Institute of Development Studies, Dhaka.

Hossain, M.S., Rahman, M.M., Harun-ur-Rashid, M., Farid, A.T.M., Quayyum, M.A., Ahmed, M., Alam, M. S., and Hussain, K.M.S.U., 2006. *Krishi Projukti Hatboi* (Handbook of Agro-technology). 4th Edition. Bangladesh Agricultural Research Institute, Gazipur, Bangladesh.

Husain, A.M.M., Hossain, M., Janaiah, A., 2001 Hybrid Rice Adoption in Bangladesh: Socio-economic Assessment of Farmers' Experiences. BRAC Research Monograph Series No. 18. BRAC, Dhaka.

Llewelyn, R.V., Williams, J.R., 1996. Nonparametric analysis of technical, pure technical, and scale efficiencies for food crop production in East Java, Indonesia. *Agricultural Economics*, 15: 113-126.

Mahmud, W., Rahman, S.H., Zohir, S., 1994. Agricultural growth through crop diversification in Bangladesh. Food Policy in Bangladesh Working Paper No. 7. International Food Policy Research Institute (IFPRI), Washington, D.C.

McDonald, J.F., Moffit, R.A. 1980. The uses of Tobit analysis. *Review of Economics and Statistics*. 61: 318 – 321.

MoA., 1989. Crop diversification program for the Fourth Five-Year Plan of Bangladesh. Agriculture Division, Ministry of Agriculture, Dhaka.

MoFA, 1962. *Pakistan Census of Agriculture – 1960. Final Report: East Pakistan.* Volume 1. Agricultural Census Organization, Ministry of Food and Agriculture, Government of Pakistan. Karachi.

Morris, M., Chowdhury, N., Meisner, C. 1996. Economics of wheat production in Bangladesh. *Food Policy*, 21: 541–560.

Negatu, W., Parikh, A. 1999. The impact of perception and other factors on the adoption of agricultural technology in the Moret and Jiru Woreda (district) of Ethiopia. *Agricultural Economics*. 21: 205 – 216.

Nkamleu, G.B., Adesina, A.A., 2000. Determinants of chemical input use in peri-urban lowland systems: bivariate probit analysis in Cameroon. *Agricultural Systems*, 63: 111–121.

PC, 1998. *The Fifth Five Year Plan (1997-2002).* Ministry of Planning, Government of Bangladesh, Dhaka.

Rahman, S. 2003. Environmental impacts of modern agricultural technology diffusion in Bangladesh: an analysis of farmers' perceptions and their determinants. *Journal of Environmental Management*. 68: 183–191.

Rahman, S. 2005. Environmental impacts of technological change in Bangladesh agriculture: farmers' perceptions, determinants and effect on resource allocation decisions. *Agricultural Economics*. 33: 107–116.

Rahman, S., Parkinson, R.J. 2007. Soil fertility and productivity relationships in rice production system, Bangladesh. *Agricultural Systems*. 92: 318-333.

Rahman, S., Thapa, G.B., 1999. Environmental impacts of technological change in Bangladesh agriculture: farmers' perceptions and empirical evidence. *Outlook on Agriculture*. 28: 233–238.

Rahman, S., 1998. Socio-economic and environmental impacts of technological change in Bangladesh agriculture. Unpublished PhD dissertation. Asian Institute of Technology, Bangkok.

Ransom, J.K., Paudyal, K., Adhikari, K., 2003. Adoption of improved maize varieties in the hills of Nepal. *Agricultural Economics*, 29: 299–305.

Shiyani, R.L., Joshi, P.K., Asokan, M., Bantilan, M.C.S., 2002. Adoption of improved chickpea varieties: KRIBHCO experience in tribal region of Gujarat, India. *Agricultural Economics*, 27: 33–39.

In: Agricultural Systems: Economics, Technology and Diversity ISBN 978-1-60692-025-1
Editor: Oliver W. Castalonge

Chapter 8

HERD MANAGEMENT, PRODUCTIVITY, AND LAND USE PATTERN OF DUAL-PURPOSE CATTLE FARMS IN THE SEMI-HUMID REGION OF CENTRAL NICARAGUA: WHAT ARE THE MAIN OBSTACLES FOR ENVIRONMENTALLY SOUND INTENSIFICATION?

Wataru Yamamoto* and Muhammad Ibrahim
[1]CATIE (Tropical Agriculture Research and Higher Education Center),
A.P. 7170 Turrialba, Costa Rica
[2]University of Wales, Bangor

ABSTRACT

Cattle production in the humid lowland tropics has been criticised for its extensive management system, low productivity and high burden on the environment. Dual-purpose cattle production systems have been preferred by small and medium sized farms in such regions due to low capital and technical demands with low risk for farmers. Intensification of existing dual-purpose cattle farms has been recognized as a key target to reduce deforestation in Central America. However, the details of complex farming systems using seasonal cattle movement as well as diversified grazing lands with pasture and naturally regenerated trees are not well known. The objective of the chapter is to examine dual-purpose cattle production systems with regard to herd management, productivity and land use pattern. Seventy-four farms in the semi-humid region of central Nicaragua were seasonally monitored and relations with land use patterns were examined.

The study results showed that grazing lands were largely covered by tree cover (23% on average). Although cattle were frequently moved from and to the farms, stocking rates did not differ by season ($P<0.05$), suggesting that seasonal cattle movement towards more humid area is limited. The maintained stocking rates as well as significantly higher occurrences of calving ($P<0.05$) seemed to cause serious fodder shortage for cattle in the

* Corresponding author: Wataru Yamamoto, Address: 49 Edgecombe Crescent, Dartmouth NS Canada B2V2A7. Tel/Fax: 1-902-433-1782. Email: wyamamoto99@hotmail.com

dry season, resulting in lower saleable milk production per cow and per hectare ($P<0.01$) and a tendency to have higher adult mortality rates. However, milk production per farm was not reduced in the dry season due to the larger number of calving. Fodder availability in the dry season significantly restricts potential of milk production. Smaller farms tend to have higher stocking rates and overgrazed, while some farms started introducing cut and carry system maintaining their pasture resources. Farms with higher financial resource improve their pasture, but shift their production types to steer rearing which requires less labour inputs.

The study concluded that shortage of dry season fodder, lack of calving control, and reluctance to increase payment for labour are the main obstacles for environmentally sound intensification of dual-purpose cattle farms. Further research is recommended on improvement of calving control, fodder availability in the dry season, the feasibility of pasture improvement, cattle movement and sharing system, and relations between farm manager and paid labour.

Key words: dual-purpose cattle production, stocking rate, calving rate, tree density, cattle movement, milk production

1. INTRODUCTION

Conversion of land use from tropical forest to pasture has been criticised as one of the causes of deforestation in the humid tropics of Central America (Parsons, 1976; Myer, 1981; Kaimowitz, 1996). At the old agricultural frontier in central Nicaragua, due to the political pressure during the civil war in 1980s, large cattle farms that had been abandoned during the civil war in 1980s were divided into small and medium sized farms in the early 1990s (Levard et al., 2001). Presently, 90% of accessible cultivated lands in lowland are occupied by pastures, and small farms comprises most of the total farms (NITLAPAN, 1995). In order to take advantage of family labour and to overcome constraints by limited land and capital, such small and medium sized farms are largely engaged in dual-purpose cattle production systems. Dual-purpose cattle production systems have been preferred by small and medium scaled farms due to the following advantages: 1) reduced risk from price changes of milk and meat, 2) higher economic productivity per area than meat production, 3) adaptation of dual-purpose cattle to the climate condition of lowland tropics, 4) less capital investment and technical support required than for specialized milk production, and 5) reduction of mastitis incidence due to calf suckling (Holmann, 1989; Souza de Abreu, 2002).

Nicaragua is considered to be one of the poorest countries in Latin America. It has a population of 5.3 million, a per capita GNI of US$720, and 48% of the population is under the national poverty line (World Bank, 2004). Cattle production is an important activity of the agricultural sector in Nicaragua. In the last few years cattle population has increased reaching more than 3.3 million head in 2001 (FAO, 2003). In particular milk export has increased since 1996 reaching US$3.5 million in 2001, which is 30 times larger than 10 years ago (FAO, 2003). In order to reduce the effects of deforestation, intensification of existing dual-purpose cattle farms was recognized as a key target in Central America (Nicholson et al., 1995).

Dual-purpose cattle production systems of lowland tropics are of intermediate intensity, are very efficient in the use of medium and poor quality forage resources, and have low calving rates and milk production per hectare per annum (Sere and De Vaccaro, 1985). The quantity and quality of products vary adjusting to the local conditions and the prices. Land and cattle constitute the main capital investment in the system and mating is natural or uncontrolled. Rotational grazing and some form of weed control is being conducted (Yamamoto, 2004). Dual-purpose cattle farms in central Nicaragua have higher stocking rates than larger farms (Ruiz, 1994) and grazing areas are largely occupied by natural pastures and naturally regenerated trees (Yamamoto, 2004). Many of these trees are beneficial for cattle production, particularly in the dry season. Simultaneously, in order to improve productivity and maintain cattle in the dry season, cultivated pastures, including those for cut and carry, were sown to a certain extent in the region (Yamamoto, 2004). On the other hand, as a feature of the dual-purpose cattle production systems, herds are composed of various types of cattle: lactating and dry cows, heifers, bulls, and calves, whose nutritional requirements are different from one another. As a results, farmers need to control pasture lands with a variety of pasture conditions and tree densities, stocking rates composed of different cattle types, and supplemental feedings within economic constraints. The objective of this paper is to assess land use patterns of grazing areas by satellite images, and herd management of dual-purpose cattle farms and to examine the relationships between them. In order to contribute to knowledge for potential modification (intensification without land degradation), the emphasis was placed on examining the level of intensification for the indicators of herd management and land use patterns by classifying the farm types.

2. Materials and Methods

2.1. Site

The study area is located in Matiguas Municipality in Matagalpa Department, central Nicaragua: Latitude 12^0 50' north and 85^0 27' Longitude east. The altitude is 200-500m above sea level. The climate of the study area is sub-humid tropical with a well-defined dry season between February and May (savanna (Aw) by Koppen climate classification). Annual rainfall varies from 1,300 to 2,000 mm, while temperature fluctuates between 28 and 32^0C. Topographically, the land is generally flat with modest slopes (0-30%) but with small areas of steep slopes (> 30 %) (INTA, 1998).

Population density of the study area is approximately 29 inhabitants/km^2 (Levard et al., 2001). Cattle production is the main economic activity in the region where pasture dominates lowland (altitude <400m) areas (Maldidier and Marchetti, 1996). The soils are generally heavy clayey, which are difficult to cultivate with light equipment and are thus more suitable for cattle production than for crop cultivation (Levard et al., 2001).

The Matiguas region is considered to be a part of "the old agricultural frontier" where large immigration occurred in the late 1940s seeking land for extensive grazing to meet the demands of the international meat market (Maldidier and Marchetti, 1996). Due to the political pressure during the civil war in the 1980s, large cattle farms were abandoned and in the early 1990s were divided into small and medium sized farms and made available to landless farmers (Levard et al., 2001).

2.2. Farm Selection

This study was based on data from the project "Regional integrated silvopastoral approaches to ecosystem management project" undertaken by CATIE and NITLAPAN under Global Environment Facility (GEF) of World Bank.

The Matiguas region is divided into four zones based on agro-ecological conditions: 1) relatively flat areas near the town of Matiguas where medium scale cattle farm enterprises are present (annual precipitation 1200-1600mm); 2) the south-eastern part of Matiguas where medium scale cattle farms are present under more humid condition (>2000mm p.a.), 3) north-western Matiguas where intensive farms are present in the highlands (>500m above the sea level); 4) the central part of the municipality and north of the town of Matiguas with steeper slope (30-50%) and moderate precipitation (1600-2000mm) (Levard et al., 2001). Sample farms were selected from two micro-watersheds: Limas and Paiwas covering three of the four agro zones (1, 2, and 4 of the classification above), which were considered to be important for cattle production. The town of Matiguas is located at the northern side of Limas. The distance between the two watersheds is approximately 20 km. Both areas are served by milk collectors.

In each watershed, initially all the farmers were invited to participate in the project. Information about the project was disseminated by project staff of NITLAPAN through visits to those farms in the NITLAPAN data base. After several workshops and farmer interviews, the farms were selected based on the criteria: (1) farm size (10-140 ha); (2) farms with more than 3 cattle; (3) willingness to participate in the project and (4) accessibility.

According to the database owned by NITLAPAN, there were approximately 190 farms in the watersheds and 130 farms were initially selected by the project at the beginning of 2003. The intention was to complete farm surveys using farm visits and project meetings. Farmers were provided with financial incentives to participate in the project. However, due to difficulties encountered when trying to meet farmers (e.g. selling or abandoning the farms, poor attendance at meetings, or high water level of the river during the rainy season) collection of both land use and herd data were completed for only 74 farms by the end of 2003.

2.3. Land Use Survey

Satellite images taken by Quick Bird (Resolution 0.7m with three natural colours) in January 2003 were used for land analysis. Farm boundaries were mapped out with farmers' participation using satellite photos based and on-site geo-referencing with GPS. After the farm mapping, boundaries of land use types inside each farm were specified in the satellite images. Land use types in the farms were classified into annual and perennial crop cultivations, pasture land, forage bank, fallows and forests based on the observations of tree density. Grazing lands were classified into nine types based on the conditions of dominant pastures and tree densities (Table 1). Pasture in the grazing areas were classified into three types: (a) degraded pasture (more than 50% are covered by herbaceous and woody broadleaves); (b) cultivated pasture (introduced pasture species under management which are generally considered as suitable for cattle production, e.g., Asia (*Panicum maximum*), Brizanta (*Brachiaria brizantha*), Jaragua (*Hyparrhenia rufa*), Estrella (*Cynodon nlemfluensis*), and Gamba (*Andropogon gayanus*); (c) natural pasture (all other pasture species including native and naturalised species).

Table 1. Land use types of grazing lands

Land use types	Abbre.[1]	Grass cover	Tree density
Degraded pasture	DGPS	< 50%	N.A.
Natural pastures with no trees	NTNP	> 50%	Nominal
Cultivated pasture with no trees	NTCP	> 50%	Nominal
Natural pasture with low tree density	LTNP	> 50%	<30 trees/ha
Cultivated pasture with low tree density	LTCP	> 50%	<30 trees/ha
Natural pasture with moderate tree density	MTNP	> 50%	>30 trees/ha
Cultivated pasture with moderate tree density	MTCP	> 50%	>30 trees/ha
Fallow	FAL	None	N.A.
Riparian forests[2]	FRST	None	N.A

N.A. Not available.

[1]These abbreviations were used for variables of the proportion of the land use types for grazing areas.

[2]Forest area excluding secondary and primary forests.

The tree density was classified into three groups: high (more than 30 trees (DBH>10cm)/hectare), low (less than 30 trees (DBH>10cm)/hectare) and none. Hereinafter the four types of land use types, natural and cultivated pastures with low and moderate tree densities are called silvopastoral areas. A tree density of 30 trees/hectare was used because tree density over 30 trees/hectare is generally considered to affect grass cover (Murgueitio et al., 2003). In addition to the pasture in the grazing areas, the supplementary forages for cut and carry [King grass (*Pennisetum purpureum x P.typhoides*), Taiwan (*Pennisetum purpureum*) and Sugar cane (*Saccharum officinarum*)] are sown in forage banks.

Cattle were grazed mainly on pasture lands, but they also had access to fallow and forests. Fallows and riparian forest were included, but secondary and primary forests were excluded from grazing areas since generally cattle did not graze in these areas. Types and conditions of pastures (natural, cultivated or degraded) and tree densities were verified during the wet season in 2003 by field observations for all land uses of all the sample farms.

In addition to the nine types of grazing areas, areas of recently sown pasture species (supplemental cut and carry forages, mainly *Pennisetum spp.*, and *Brachiaria brizantha*) were carefully considered since these pasture species may have a significant influence on cattle production. Hence, the proportions of these two types of pasture within grazing areas were recorded.

2.4. Tree Cover Study

Tree cover was recorded for each land use type. Estimation of tree cover was carried out on the images with the 1:5,000 scale since one tree cover as large as 56 m^2 (equal to the size of 7.5m^2, approximately 8.5m in diameter of trees) is seen as a point. The individual trees smaller than this size were ignored. Tree cover was analysed by ARC VIEW 3.3 as follows.

1) Forest areas where lands were totally covered by trees were manually drawn on the satellite image for each farm.
2) On the images with a 1:5000 scale, large dispersed trees as large as 56 m^2 (equal to the size of 7.5m^2, approximately 8.5 m in diameter of tree cover) or groups of trees were manually replaced by points based on visually estimation by placing the number of points in proportion to the size of the tree cover. Thus, one point was equivalent to the size of 7.5 m square on the images, which was approximately 8.5 m in diameter of trees.
3) The points placed (one 7.5m^2 on the images) were converted into nine 2.5m^2 pixels (2.5m interval grid on 7.5 m^2) so that one tree cover in more than two land use types were counted for each land use type by dividing the tree cover by 2.5m^2 square pixels.

2.5. Herd Survey

Data for herd size and structures, as well as changes in herd inventory (number of animals sold/bought and born/died), daily saleable milk yields (milk produced minus calf suckling and family consumption) and supplementary feeding were collected by structured farmer interviews. Farmers were asked for information about the actual number of cattle on the farm at the time of the interviews and occurrences of calving, mortality, animal sales and purchases in the previous three months (i.e. since the last interview). Cattle were classified into lactating and dry cows, heifers, bulls and oxen (castrated male cattle for draft purpose), steers in the fattening stage (male cattle older than 2.5 years old), steers in the rearing stage (male cattle between 1.5 and 2.5 years old), and calves (before and after weaning).

The interviews were conducted for 74 farms every 3 months for one year period in 2003: 1) between the end of January and the beginning of February (the beginning of the dry season), the end of April and the beginning of May (the end of the dry season), the end of July and the beginning of August (the beginning of the wet season) and the end of October and in the beginning of November (the end of the wet season), (hereinafter Jan./Feb., Apr./May, Jul./Aug. and Oct./Nov.). During the survey it was observed that some farmers did not correctly explain the number of cattle on the farm due to the complexities of their herd management: for example, confusion between the cattle owned by the farmers and the cattle kept on the farm. Therefore, in the last set of interviews, farmers were specifically asked regarding overall herd movement during the one year period and herd size was subsequently adjusted by removing or adding the number of cattle which were or were not on the farm for a certain period. Also, in some cases farmers had additional lands where their cattle can graze, but these areas were outside the project area. Stocking rates were adjusted by adding the additional lands to the farms.

2.6. Data Analysis

The data were analysed by using MINITAB 13 (MINITAB, 2000) using correlation analysis to examine the relationships between land use and herd management variables. One way analysis of variance were used for statistical analysis for stocking rates and milk production per hectare since they were normally distributed. Kruskal-Wallis tests were used for statistical analysis for all other data, both non-parametric data and non-normally distributed parametric data.

2.6.1. Land Use Types and Tree Cover

Mean areas and percentages of each land use type were calculated and presented for entire farm lands and for grazing areas by the classification shown in Table 1. Mean tree cover for the land use types of grazing areas were also calculated. The percentages of the land use types for grazing areas were compared by the size of grazing areas (<15ha, 15-30ha and >30ha).

2.6.2. Herd Size and Structure

Herd sizes were estimated based on livestock units (LU) (one livestock unit is equivalent to 400kg of liveweight). As weighing animals were not possible, the following equivalencies were used: 1.0 for lactating and dry cows, 0.75 for heifers (1.5-3 years), 1.0 for steers in the fattening stage (older than 3 years), 1.25 for bulls and oxen, 0.75 for steers in the rearing stage (1.5-3 years old), 0.5 for weaned calves, and 0.25 for calves before weaning. Herd sizes and compositions were calculated for four seasons with adjustment for cattle movement using the following equation.

Annual mean herd size (LU) = $1/4 \times \Sigma^{4}_{i=1}$ *Herd* (i=Seasons,1,2,3,4)

where Herd $_{1,2,3,4}$ (LU) = 3 (months) x Actual number of cattle in the farm (LU)
- Number of cattle removed (LU) x Removed period (month)
+ Number of cattle placed (LU) x Placed period (month)

2.6.3. Stocking Rates

Stocking rates were estimated by dividing an total herd size (LU basis) by the estimated grazing area (pasture lands, fallows and riparian forests) based on satellite images. Where farmers used additional land used for grazing cattle, the stocking rates were adjusted by adding the additional land to the farms (8 farms, 5.3% of total grazing areas added). Calculations were undertaken for four seasons and the means of the four values were used as the annual stocking rates. The formula used to estimate stocking rate was as follows.

$$\text{Stocking rate} = \frac{\text{Herd size (LU)}}{\text{Grazing areas (hectare)} + \text{Additional grazing areas (hectare)}}$$

2.6.4. Calving and Mortality Rates

Seasonal and annual calving and mortality were monitored and compared by season. These were calculated by the following equations:

Annual/seasonal calving rates = (Frequency of calving through the year or of the period) / (Annual mean or seasonal total number of lactating and dry cows) x 100

Annual/seasonal adult mortality rates = (adults died throughout the year or of the period) / (annual mean or seasonal total number of adults) x 100

Annual/seasonal calf mortality rates = (calves died throughout the year or the period) / (total number of calvings throughout the year or the period) x 100

2.6.5. Cattle Sales and Purchase

Sales and purchases of all categories of cattle, and daily saleable milk production per farm, per lactating cow and per hectare of grazing areas were calculated and statistically compared by season. Annual/seasonal sales and purchases of the cattle were presented by the percentages of the animal type (cow, steer and calf) sold or purchased in relation to the annual means or seasonal number of the animal type. Total annual income from animal and milk sales was calculated based on the mean prices obtained in the general farm survey (Yamamoto, 2004).

2.6.6. Analysis of Supplementary Feeding

The sample farms were classified based on the use of supplementary feeds including salt, molasses, supplementary forages by cut and carry, and concentrates. The same variables (except for the percentage of area with moderate tree density) used for the analysis of cattle movement were applied to analyse the effects of the type of supplements used in the sample farms.

2.6.7. Correlation Analysis

Correlations were determined between land use parameters, size of grazing areas and herd parameters. The selected variables for the analysis were land use parameters (proportion of the land use types for entire grazing area), herd parameters [stocking rates (LU/ha); proportion of both milking and dry cows, milking cows and steers for the herds (LU base); and mean daily saleable milk production per hectare of grazing area (litres/ha/day); and annual income from cattle production (US$/ha/year)], and pasture parameters (proportion of recently sown areas with *B. brizantha* and cut and carry forages in relation to the entire grazing areas).

3. RESULTS

3.1. Land Use and Tree Cover

On average, farm size was 26.5 hectares of which grazing lands occupied approximately 80% (Table 2). Two thirds of grazing areas were occupied by degraded pasture (32 %), natural pasture with moderate tree density (17%), and riparian forests (17%), which have significant tree cover (Table 3). Pastures with no trees occupied only 3.3% of the grazing areas. Areas occupied by cultivated pastures were smaller than those of natural pastures (14.8% against 29.9%). As a whole, the results suggest that grazing areas were largely occupied by silvopastoral areas where trees were present with pastures, and natural and degraded pastures were more common than cultivated pastures.

Table 2. Mean land use of dual-purpose cattle farms in the Matiguas region, central Nicaragua

Land use type	Area (ha)	Standard error	Percentage (%)
Annual crop cultivation	2.0	7.67	7.7
Perennial crop cultivation	0.5	2.0	2.0
Forage bank	0.8	3.0	3.0
Pasture lands[1]	16.4	1.42	61.8
Fallow	1.4	0.28	5.1
Forest[2]	5.4	0.66	20.2
Others	0.1	0.02	0.2
Grazing area[3]	21.3	2.00	80.0
Mean per farm	26.5	2.06	100.0

Source: Satellite images based on Quick Bird, January 2003, n=74.
[1]Pasture lands were classified into seven land use types (Table 1).
[2]Forest is composed of riparian forest, secondary and primary forests.
[3]Grazing area is composed of seven land use types of pasture lands, fallow and riparian forest.

Table 3. Description of land use types of grazing areas in the Matiguas region, central Nicaragua

Land use type	Abbrev.	Mean (ha)	Standard error (ha)	Percentage for grazing area (%)	Tree cover (%)
Grazing areas					
Degraded pasture[1]	DGPS	6.9	1.15	32.2	18.8
Natural pasture with no trees[1]	NTNP	0.5	0.17	2.4	7.0
Cultivated pasture with no trees[1]	NTCP	0.2	0.09	0.9	4.6
Natural pasture with low tree density[1]	LTNP	2.2	0.36	10.4	12.1
Cultivated pasture with low tree density[1]	LTCP	1.5	0.33	7.0	12.5
Natural pasture with moderate tree density[1]	MTNP	3.7	0.57	17.2	24.8
Cultivated pasture with moderate tree density[1]	MTCP	1.5	0.40	6.9	19.6
Fallow[1 2]	FAL	1.4	0.28	6.4	42.6
Riparian forest[1 2]	FRST	3.5	13.52	16.6	84.8
Total		21.3	2.06	100.0	23.2
Land use types of specific pasture species					
Cut and carry forages[4]	CCF	0.8	0.15	3.0	
B.brizantha[5]	BB	0.4	0.14	1.6	

Source: Satellite images based on Quick Bird, January 2003, n=74.
[1]These land use types form grazing areas.
[2]These land use types were not included in pasture lands, but included in grazing areas.
[3]Tree cover was calculated by estimation at 1:5,000 scale.
[4] Outside of grazing areas (equivalent to Forage bank in Table 2).
[5] Included in grazing areas (within cultivated pastures).

Grazing areas had 23% tree cover on average (Table 3). It should be noted that degraded pasture had 18.8% tree cover which was close to the range of moderate tree density (19.6% for cultivated pasture with moderate tree density).

3.2. Land Use and Farm Size

The comparison of land use pattern by farm size show that larger farms tended to have more degraded pasture, but less natural pasture with low tree density (Table 4). Cultivated pasture with no trees (NTCP) and moderate tree density (MTCP) were larger on medium sized farms ($P<0.05$ for MTCP between <15ha and 15-30ha, tendency for NTCP), but those with low tree density did not differ by farm size. Cultivated pasture with no trees showed high standard errors for all farm types, suggesting that introduction of intensive pasture management is not related to farm size.

Table 4. Percentages of land use types by size of grazing area in the Matiguas region, central Nicaragua

Land use types	Category Abbrev.	< 15 ha	15-30 ha	> 30 ha
Degraded pasture	DGPS	21.8±4.1	30.0±6.1	37.7±7.3
Natural pasture with no trees	NTNP	1.6±0.7	1.8±0.9	2.9±1.4
Cultivated pasture with no trees	NTCP	0.3±0.2	1.6±1.0	0.8±0.7
Natural pasture with low tree density	LTNP	18.0±3.0	10.7±2.9	6.9±2.8
Cultivated pasture with low tree density	LTCP	7.7±2.7	8.4±2.6	5.8±2.4
Natural pasture with moderate tree density	MTNP	20.5±3.6	16.3±4.6	16.2±4.7
Cultivated pasture with moderate tree density	MTCP	5.4±1.6	11.5±3.3	5.7±4.4
Fallow[1]	FAL	10.8±3.6	4.3±1.6	5.9±2.1
Riparian forests[1]	FRST	13.9±1.8	15.3±1.8	18.2±2.1
Total	Total	100.0	100.0	100.0

Note: Means ±Standard errors, Unit : %.

3.3. Herd Size and Structure

The results of herd structuring monitoring show that stocking rates and herd sizes did not differ by season ($P<0.05$, Table 5). Herd size was at its largest at the beginning of the dry season and lowest at the end of the dry season. In fact, stocking rates tended to be lower at the beginning of the wet season than the end of the dry season, but herd size tended to be at its lowest at the end of dry season, indicating that cattle were concentrated in larger farms in the beginning of the wet season.

The numbers of lactating cows in the herds increased toward the end of the dry season and the beginning of the wet season. The number of dry cows and steers in the fattening stage differed between seasons ($P<0.05$ for dry cows and $P<0.01$ for steers in fattening stages).

Table 5. Herd structure and stocking rate for each season

Season	Month	Lactating cows	Dry cows	Heifers	Steers under rearing	Steers under fattening	Oxen	Calves	Total [1]	Stocking rate[1]
Dry	Jan./Feb	7.5 ± 0.74	5.2 ± 0.57	3.3 ± 0.68	1.3 ± 0.66	0.5 ± 0.28	0.2±0.06	8.4±0.93	19.8±1.91	0.95±0.06
	Apr./May	8.0 ±0.93	3.6 ± 0.48	3.4 ± 0.63	0.9 ± 0.38	0	0.2±0.07	9.7±1.10	18.8±1.89	0.90±0.05
Wet	July/Aug	8.4 ± 0.83	3.3 ± 0.47	3.3 ± 0.71	1.2 ± 0.68	0.1 ± 0.12	0.2±0.08	9.3±0.88	18.9±1.78	0.87±0.04
	Oct./Nov.	7.3 ± 0.83	4.4 ± 0.47	4.2 ± 0.71	1.6 ± 0.68	0	0.2±0.08	8.8±0.88	19.7±1.90	0.93±0.05
Annual Mean		7.81 ± 0.73	4.11 ± 0.43	3.55 ± 0.57	1.25 ± 0.51	0.15 ± 0.07	0.20±0.07	9.04±0.81	19.3±1.68	0.92±0.04

Note: Means ±Standard Errors. n=74, Unit: number of animals per farm (except for Total and Stocking rates by livestock unit.

Table 6. Calving and mortality for each season

Season	Month	Calving (% of cow)	Adult mortality (% of adults)	Calf mortality (% of calvings)
Dry	Jan./Feb.	14.5 ±1.7 (31)	0	4.6 ±1.8 (29)
	Apr./May	15.8 ±2.6 (34)	0.57±0.30 (57)	1.5 ±0.7 (9)
Wet	July/Aug.	8.0±1.9 (17)	0.21±0.20 (21)	3.0 ±1.7 (19)
	Oct./Nov.	7.9 ±1.6 (17)	0.21±0.16 (21)	6.1 ±3.1 (39)
Annual mean		47.0±4.1	1.0±0.36	15.8±4.2

Note: Means ± Standard Errors (% of annual mean), unit: %, n=74.

Calculated by the number of occurrences in relations to annual mean and seasonal number of the animal type.

The results of the comparison between seasons suggested that the number of dry cows was significantly higher at the beginning of the dry season than at the end of the dry season ($P<0.01$) and the beginning of the wet season ($P<0.01$). The number of steers in the fattening stages at the beginning of dry season was greater than at the end of the dry season ($P<0.05$) and the end of the wet season ($P<0.05$). The results suggest that herd sizes were influenced by dry cows and steers in the fattening stage (at some farms), but they did not significantly change the stocking rates of the entire farm.

3.4. Calving and Mortality

The study results show that a greater number of calves were born in the dry season (Jan/Feb and Apr/May) than in the wet season (July/Aug and Oct/Nov) ($P<0.01$) (Table 6). Whilst the percentages of calf mortality was higher in the end of the wet season, the percentages of adult mortality tended to be higher at the end of the dry season. The mean annual calving rate, adult death rate and calf death rate (±SE) were 47.0±4.13 %, 1.0±0.36 % and 15.8±4.23 % respectively.

3.5. Sales and Purchases of Cattle

The results of farm monitoring show that calf/steer sales and calf/cow/heifer purchases did not differ by season ($P<0.05$), while cow sales ($P<0.05$) and steer purchases ($P<0.05$) differed by season (Table 7). Higher percentages of cows were sold at the end of the dry season than the beginning of dry season ($P<0.01$), and at the beginning of wet season than the beginning of dry season ($P<0.05$), while steers were purchased only at the beginning of dry season. Steer sales and purchase were reported from only four and six farms respectively. On average, 11% of cows and 19% of calves were sold annually.

Table 7. Cattle sales and purchases for each animal type for each season

Season	Month	Sales			Purchases		
		Cows	Steers	Calves	Cow/Heifer	Steers	Calves
Dry	Jan./Feb.	1.9+0.9	0.2±0.2	5.2±1.4	0.5±0.3	5.2±2.3	0.3±0.3
	Apr./May	5.6+1.5	2.5±1.8	7.1±1.9	0.4±0.2	0	3.4±1.7
Wet	July/Aug	2.9+0.7	0	5.3±1.2	0.5±0.2	0	5.3±2.4
	Oct./Nov.	3.9+1.6	0	6.8±2.0	3.3±1.6	0	1.9±1.4
Annual mean[1]		10.8+1.8	4.2±2.2	19.1±2.6	4.5±1.7	6.6±2.9	10.1±3.7

Note: Means ± Standard Errors, Unit: %, n=74.

[1]Number of occurrences divided by mean number of the cattle types.

3.6. Milk Production

The study results show that saleable milk production differed significantly between seasons on a per cow (P<0.01) and per hectare (P<0.05) basis, but did not differ on a per farm basis (Table 8).

Table 8. Daily saleable milk production for each season

Season	Dry		Wet		
Category/Month	Jan./Feb.	Apr./May	Jul./Aug.	Oct./Nov.	Annual Mean
Per farm (litres/farm)	28.0 ± 3.2	22.8 ± 2.8	33.2 ± 3.4	25.6± 2.7	27.4± 2.7
Per lactating cow (litres/cow)	3.33 ± 0.18	2.70 ± 0.17	3.83 ± 0.15	3.56± 0.18	3.35± 0.10
Per hectare (litres/ha)	1.33 ± 0.11	1.20 ± 0.09	1.65 ± 0.13	1.35± 0.12	1.38± 0.08

Note: Means ± Standard Errors, n=74.

Saleable milk production per cow was significantly lower at the end of the dry season than other seasons (P<0.01) and it was also lower at the beginning of the dry season than the beginning of the wet season (P<0.01). It should be noted that saleable milk production per cow at the end of the dry season was not only low on average but was also extremely low on some farms.

Saleable milk production per hectare at the end of the dry season was significantly lower than the beginning of the wet season (P<0.01). Saleable milk production per hectare did not change between the end of the wet season and the end of the dry season (P<0.05).

Saleable milk production per farm tended to be at its lowest level at the end of the dry season and highest at the beginning of the wet season. It should be noted that daily saleable milk production per farm was likely to be higher at the beginning of the dry season than the end of wet season though milk yield per cow was lower at the beginning of dry season, indicating that more lactating cows were present in the farms in the dry season. Moreover, estimated milk yield per cow, excluding the farms with no saleable milk production were 3.62±0.15 litres/cow (n=67), 3.08±0.15 litres/cow (n=65), 4.05±0.12 litres/cow (n=70), and 3.88±0.14 litres/cow (n=68) for Jan./Feb. Apr./May, Jul./Aug., and Oct./Nov. respectively.

3.7. Annual Income from Cattle Production

The study results showed that on average milk contributed to 73% of the income on the farms (Table 9). On average sales of cows and calves contributed 51% and 44% of the total income from animal sales. Cow and calf sales were reported by approximately half of the sample farms (36 and 40 farms respectively), but steer sales were reported by only 4 farms, resulting in large standard errors.

Table 9. Total annual farm income from cattle production

Category		Animals				Milk	Total
	Unit	Cow	Steer	Calf	Total		
Value	US$	410±80	39±27	353±73	802±128	2,179±218	2,981+300
Percentage	%	13.7	1.3	11.9	26.9	73.1	100

Note: Means ± Standard errors. N= 74.

Based on mean numbers of cattle and amount of milk sold and producer prices (Yamamoto, 2004): milk price U$0.19/litre in the wet season, US$ 0.25/litre in the dry season, cow (US$228), steer (US$182), calf (US$105).

3.8. Supplementary Feeding

Sample farms are classified into six categories on the basis of supplementary feeding. Size of grazing areas, stocking rates, and percentage of degraded/cultivated pastures are compared in Table 10. All the farms used supplementary feeding for cattle. Two thirds of the farmers used salt with minerals, half of the farms used supplementary forages by cut and carry system. The use of concentrates was limited (5.4% of the sample farms). The amount of cut and carry forages offered to animals varied among the farms from less than 4 to 32 kg/head/day depending upon the availability and number of cattle. The supplementary pastures were minced manually or by machine depending upon the availability of machinery and labour. Some farmers fed entire plants to cattle. The amount of molasses used also varied, but most of the farmers gave a very limited amount (approximately 1-2 litres/day/farm). Farms which only used salt as a supplement and farms which used supplementary pastures represented 33.8% and 64.8% of the total, respectively. Size of grazing areas, stocking rates, and proportion of degraded/cultivated pastures did not differ significantly by farm types based on supplementary feeding (P<0.05).

Table 10. Characteristics of grazing areas and stocking rates by farm type based on supplementary feeding

Farm types	No.	%	Grazing area ha	Stocking rates LU/ha	Degraded pastures %	Cultivated pastures %
Natural/mineral salt only	25	33.8	21.0±2.9	0.80±0.06	20.3±3.8	21.9±5.0
Salt and supplementary pastures	14	18.9	19.6±3.8	0.89±0.09	31.7±8.5	14.1±5.4
Salt, supplementary pastures and molasses	30	40.5	21.0±2.9	0.95±0.07	29.7±5.4	13.5±3.3
Salt and molasses without supplementary pasture	1	1.4	27.0	0.72	6.97	16.9
Mineral salt, supplementary pasture and concentrates	4	5.4	52.4±17.1	0.92±0.11	51.1±17.4	2.7±2.0
Total	74	100				

Note: Means ±Standard Errors. Fr. Frequency.

4. Correlation Analysis

4.1. Correlations between Proportions of Land Use Parameters

The results of correlations between size of grazing areas and land use parameters show negative correlations between size of grazing areas and the proportion of natural pastures with low tree density (LTNP) ($P<0.05$) and significant negative correlations between degraded pasture (DGPS) and many land use types (DGPS-NTNP $P<0.05$, DGPS-LTNP $P<0.001$, DGPS-MTNP $P<0.001$, DGPS-LTCP $P<0.05$) (Table 11). The correlations between no trees and low tree density of both natural and cultivated pastures were significant (NTNP-LTNP, $P<0.001$ and NTCP-LTCP, $P<0.01$), suggesting that these land uses were developed in a similar pattern in the sample farms. However, cultivated pasture with moderate tree density (MTCP) was not significantly correlated with any other land use types, indicating that this land use type was independently developed in the farms. Regarding pasture variables, cut and carry forages and LTNP (CCF-LTNP $r=0.26$ $P<0.05$) as well as *B. brizantha* and MTCP (BB-MTCP $P<0.05$) were significantly correlated.

4.2. Correlations between Size of Grazing Areas and Herd Parameters

The results of correlations between size of grazing areas and herd parameters showed that size of grazing areas was highly correlated with herd size ($P<0.01$), but not with other herd parameters (Table 12). The proportions of cows and steers were negatively correlated ($P<0.001$), indicating that these cattle types substituted for one another. In addition, the correlation between the proportion of cows and lactating cows for herds were positive ($P<0.001$). Stocking rate was correlated with both annual income from cattle production per hectare ($P<0.001$) and milk production per hectare ($P<0.001$), while the proportion of milking cows was correlated only with milk production ($P<0.001$), not with annual income.

4.3. Correlations between Land Use and Herd Parameters

The results of correlation analysis between land use, pasture and herd parameters showed that correlations were significant between stocking rates and the proportion of natural pastures with low tree density (SR-LTNP, $P<0.01$), between cultivated pasture with low tree density and the proportion of steers (LTCP-STEER, $P<0.01$), and between the proportion of cut and carry forages and saleable milk production/stocking rate (CCF-MEMLK, $P<0.01$; CCF-SR, $P<0.01$) (Table 13). In addition, farms with higher stocking rates tended to have less degraded pasture ($P=0.086$). The results suggest that farms with higher stocking rates had more natural pasture with low tree density ($P<0.01$) and cut and carry forages ($P<0.01$), and they tended to have less degraded pasture. In contrast, farms with more steers had more cultivated pasture with low tree density ($P<0.01$).

Table 11. Correlation matrix between size of grazing areas, land use and pasture parameters

Variables	Grazing area	DGPS	NTNP	NTCP	LTNP	LTCP	MTNP	MTCP	FAL	FRST	CCF
DGPS[1]	0.194 (0.098)										
NTNP[1]	0.097 (0.412)	-0.279* (0.016)									
NTCP[1]	0.031 (0.795)	-0.117 (0.321)	-0.114 (0.333)								
LTNP[1]	-0.277** (0.017)	-0.509** (0.000)	0.400** (0.000)	-0.081 (0.493)							
LTCP[1]	-0.069 (0.558)	-0.242* (0.037)	-0.111 (0.347)	0.322** (0.005)	-0.144 (0.221)						
MTNP[1]	-0.039 (0.740)	-0.505** (0.000)	0.018 (0.879)	-0.051 (0.666)	0.157 (0.183)	-0.197 (0.093)					
MTCP[1]	0.014 (0.907)	-0.194 (0.097)	-0.119 (0.314)	0.128 (0.279)	-0.173 (0.141)	0.007 (0.953)	-0.141 (0.232)				
FAL[1]	-0.129 (0.273)	-0.125 (0.289)	0.041 (0.726)	-0.073 (0.537)	-0.150 (0.202)	-0.052 (0.658)	-0.174 (0.139)	-0.178 (0.129)			
FRST[1]	0.240** (0.039)	0.090 (0.444)	-0.057 (0.630)	-0.224 (0.055)	0.001 (0.992)	-0.050 (0.671)	-0.225 (0.053)	0.018 (0.881)	-0.358** (0.002)		
CCF[1]	-0.047 (0.691)	-0.023 (0.847)	0.086 (0.468)	-0.101 (0.391)	0.263* (0.024)	-0.112 (0.341)	-0.124 (0.294)	0.294 (0.734)	-0.057 (0.627)	0.212 (0.070)	
BB[1]	0.039 (0.740)	-0.183 (0.118)	-0.097 (0.410)	-0.101 (0.394)	0.191 (0.103)	-0.026 (0.823)	-0.045 (0.704)	0.253* (0.030)	-0.060 (0.613)	0.161 (0.172)	0.070 (0.551)

Note: n=74, Cell Contents: Pearson correlation/(P-Value). * P<0.05, ** P<0.01, *** P<0.001, Significant correlations are presented in bold.

[1]Land use and pasture parameters are based on the proportions for grazing areas.

Table 12. Correlation matrix between farm size and herd parameters (n=74)

Variables	Grazing area	Herd size	Stocking rate	COW	STEER	MCOW	CTL/Y/H
/unit	ha	LU	LU/ha				US$/year/ha
Herd size	0.856*** (<0.001)						
Stocking rate	-0.210 (0.072)	0.221 (0.059)					
COW	-0.148 (0.208)	-0.157 (0.182)	-0.108 (0.359)				
STEER	0.176 (0.133)	0.140 (0.233)	-0.003 (0.980)	-0.575***(<0.001)			
MCOW	-0.154 (0.191)	-0.138 (0.240)	-0.007 (0.955)	0.667***(<0.001)	-0.435***(<0.001)		
CTL/Y/H	-0.175 (0.135)	0,067 (0.569)	0.468 (<0.001)	0.055 (0.641)	-0.059 (0.619)	0.121 (0.305)	
Milk production (litres/ha/day)	-0.235 (0.044)	0.086 (0.464)	0.642 (<0.001)	0.263 (0.023)	-0.217 (0.063)	0.514 (<0.001)	0.682 (<0.001)

Note: above: Pearson correlation, (P-Value) below.

*P<0.05, **P<0.01, ***P<0.001, Significant correlations are presented in bold.

COW: proportion of cows for herd, MCOW: proportion of milking cow for herd,

STEER: proportion of steers for herd, CTL/Y/H: annual income from cattle production per hectare

Table 13. Correlation matrix between land use and herd parameters (n=74)

Land use type	Annual income	Daily milk production	Stocking rate	Milking cow/herd	Steer / herd
Unit	US$/ha	litres/day/ha	LU/ha		
Abbrev.	CTL/Y/H	MEMLK/D/H	SR	MCOW	STEER
DGPS	0.045 (0.701)	-0.015 (0.899)	-0.201 (0.086)	0.001 (0.990)	0.013 (0.915)
NTNP	-0.170 (0.147)	-0.039 (0.739)	0.177 (0.132)	-0.126 (0.284)	0.162 (0.168)
NTCP	-0.067 (0.569)	-0.016 (0.892)	-0.168 (0.152)	0.093 (0.433)	-0.080 (0.497)
LTNP	-0.002 (0.985)	0.082 (0.488)	0.356** (0.002)	0.048 (0.687)	-0.051 (0.669)
LTCP	0.037 (0.756)	-0.066 (0.574)	-0.042 (0.724)	-0.140 (0.233)	0.362** (0.002)
MTNP	0.010 (0.933)	0.040 (0.733)	0.110 (0.350)	-0.044 (0.710)	-0.130 (0.268)
MTCP	-0.015 (0.896)	-0.003 (0.979)	-0.167 (0.156)	0.082 (0.488)	-0.093 (0.429)
FAL	-0.035 (0.767)	0.069 (0.560)	0.020 (0.862)	0.125 (0.288)	-0.119 (0.313)
FRST	-0.023 (0.847)	-0.179 (0.128)	-0.019 (0.871)	-0.089 (0.452)	0.103 (0.381)
CCF	0.223 (0.057)	0.336** (0.003)	0.391** (0.001)	0.195 (0.096)	-0.050 (0.671)
BB	0.048 (0.682)	0.164 (0.163)	0.026 (0.828)	0.141 (0.232)	0.134 (0.256)

Cell Contents: Pearson correlation (P-Value),*P<0.05, ** P<0.01. Significant correlations are presented in bold.

5. Discussion

5.1. Tree Cover, Stocking Rates, and Land Use Types

The results show that on average tree cover represented 23% of the grazing lands (Table 3). Pasture occupied 62%, while forest occupied 20% of the total farm area (Table 2). The most frequent land use types (degraded pasture (DGPS) and natural pasture with moderate tree density (MTNP)), occupied almost half of the grazing areas with moderate tree cover accounting for approximately 20%, while pastures with no trees occupied only 12% of the grazing areas. It suggests that the farms largely utilize grazing lands with trees (silvopastoral areas) for cattle grazing. Stocking rates were 0.91 on average, (Table 5) and pastures were largely composed of degraded and natural pastures rather than cultivated pastures. This suggests that the farms are under extensive management.

Correlations were positively significant between stocking rates and natural pasture with low tree density (LTNP) (P<0.01) and close to being significant between stocking rates and degraded pasture (DGPS) (P=0.086) (Table 13). These suggest that higher stocking rates tend to reduce degraded pasture and increase lower tree density by disturbing natural regeneration of trees. In fact it was proven that degraded pasture positively contributes to milk production but natural pasture with low tree density negatively contribute to milk production (Yamamoto, et al., 2007). It suggests that increasing stocking rate in sample farms causes overgrazing; the majority of farmers simply want to increase number animals in their farms (Yamamoto, 2004). Since farms with higher stocking rates were rather small (P=0.072, Table 12), these farms may not have sufficient financial resources for proper management. However, farms with higher stocking rates have more cut and carry forages (P<0.01, Table 13) and farms with more cut and carry forages has less natural pasture with low tree density (P<0.05, Table 11). It clearly suggests that some farms intensify milk production by introducing cut and carry forages, avoiding to increase less productive pasture type.

5.2. Herd Structure and Farm Types

Regarding herd structure, the proportion of cows and steers were negatively correlated (P<0.01) and the correlations between the proportion of milking cows for the entire herd and the proportions of both milking and dry cows were significant (P<0.001) (Table 12). This suggests that there are two types of farms: ones that rear steers and ones with more emphasis on milk production. Saleable milk production generates daily cash for living which is essential for farmers, but steer production generates income in the medium-term. The proportion of steers was positively correlated with the proportion of cultivated pasture with low tree density (P<0.01, Table 13), suggesting that farmers who concentrated on rearing steers had greater financial resources. They were able to sell the steers at the end of dry season when cattle prices were at their highest.

It may indicate that farms with greater financial resources invest on pasture improvement but more emphasis on rearing steers which is less labour intensive. These farmers do not intensify cattle production by increasing milk production, but shift to steer rearing potentially because they are reluctant to hire labour since it is difficult to develop reliable relationship in this culture (Kaimowitz, 1996). It is likely that farm owners who work at their farms introduced cut and carry systems (intensifying milk production), but those who do not work at their farms shifted the production types (from milk production to steer rearing).

5.3. Calving and Cattle Mortality

The study results showed that calving frequencies were significantly higher in the dry season than in the wet season (P<0.01, Table 6) and that saleable milk production per cow was significantly lower at the end of the dry season than in other seasons (P<0.01, Table 8). Also adult death rates were higher at the end of the dry season (Table 6). Considering the fact that one third of the farms did not use any supplementary fodder in the dry season (Table 10), it may suggest that stocking rates at the end of dry season are higher than the carrying capacity of the grazing areas thus causing low milk yield per cow and high adult death rates.

The average calving rate (47%) is low compared with the national census of Costa Rica, 64% (Corporacion Ganadera, 2000) and with the ranges presented by Sere and de Vaccaro (1985) (50-70%) in Honduras, Panama and Colombia, but similar to the national average of Venezuela (45%, Plasse, 1992). Ideally, calving interval should be 12 months with a dry period of two months (100% calving rate) (Ramirez, 2002), which is higher than any of these ranges. In the study area, calving rates may be generally lowered due to malnutrition during the dry season, presence of calves and low availability of bulls particularly in small farms. It was possible that calving records were not reported correctly due to difficulties associated with specifying frequencies for three months periods and because some farmers simply brought milking cows after calving to the sample farms which are conveniently located to sell milk to milk collectors. It is also possible that low calving rates were caused by the farmers' tendency to keep unproductive cows (Yamamoto, 2004). Since the variation in milk production between individual cows is high in dual-purpose cattle farms, frequent culling was suggested by Llamozas and Vaccaro (2002). On average, calf mortality rate was 16%, which is higher than the results presented by Sere and de Vaccaro (1985) and Vacarro et al. (1992) (6-12%) in Brazil, Costa Rica and Panama, and Venezuela. The average adult mortality rate was 1.0%, which is relatively low compared with the range presented by Sere and de Vaccaro (1985). Further studies are recommended on the improvement of calving rates, the timing of calving, and the cause of the death (e.g. nutritional problems, disease or accidents).

5.4. Milk Production

It is to be expected that milk production is lower in the dry season. However, the results show that daily saleable milk production did not differ per farm by season ($P<0.05$, Table 9). The number of calvings were higher in the dry season ($P<0.01$) and the number of milking cows did not differ by season ($P<0.05$) (Table 5 and 6). It is probably because more cows conceive at the beginning of the wet season when physical conditions are improved by better fodder. It seems that since most of the farms were at locations accessible to milk collectors, they tended to bring more milking cow to the farms to sell milk.

In contrast, saleable milk production by season differed significantly per cow ($P<0.01$) and per hectare ($P<0.05$) (Table 8). Clearly the conditions of grazing areas in the dry season restrict milk yield. Common pasture species in the study area, such as *P. notatum* and *H.rufa,* which covered large grazing areas, do not maintain their nutritional values in the dry season (Peters et al., 2002). This could cause low milk yields per cow and per hectare. Since the higher number of calvings occurred during the dry season ($P<0.01$, Table 6), milking cows which were intentionally kept in the farms in the dry season may also have caused low milk yield per cow and per hectare. Clearly, greater demand by calves, as a consequence of the higher number of calvings, also caused the lower saleable milk production per cow in the dry season. Further studies are recommended to consider fodder availability in the dry season as well as the feasibility and effects of pasture improvement.

Based on the data of seasonal saleable milk production per cow (Table 8), it can be estimated from all sample farms that true saleable milk production was 804 litres/cow/lactation [daily estimated milk yields per cow, 3.35 litres/day/cow multiplied by 8 month lactation period (Lavard et al., 2001)]. This yield result is similar to those presented by Sere and de Vaccaro (1985) in Colombia and Panama (749 and 956 litres/cow/lactation), Vaccaro et al., (1992) in Venezuela (700-1,100 litres/cow/lactation), and by Wilkins et al. (1979) in Bolivia (600-1,200 litres/cow/year).

Tropical grasses in general have poorer feeding value than temperate grasses due to lower voluntary intake and dry matter digestibility (Minson, 1981). Stobbs and Thompson (1975) concluded that by using improved tropical pasture it was possible to produce 2,000kg of milk per lactation. The sample farms were family farms engaged in dual-purpose cattle production, thus milk is partly consumed by calves and family members. Wilkins et al. (1979) estimated that maximum milk production per cow with improved pastures without concentrates was approximately 2,000 litres (1,300 litres sold per lactation with 700 litres consumed by a calf). This target yield is comparatively higher than the results of the present study. The grazing areas in the study area are highly covered by natural pastures (30% Table 4), which have limited potential for cattle production in the tropics (Humphreys, 1987). Stobbs and Thompson (1975) suggested that "the early lactation stage is the most important phase of the whole cycle: the cow's potential is being established". However, in the study area, the high potential may not be achieved due to the fodder restriction in the dry season when the majority of calvings occur. Based on these findings, the study concludes that there is considerable potential to increase saleable milk production per cow and per hectare in the study area.

5.5. Stocking Rates and Cattle Movement

It is generally considered that in the semi-humid tropics, farmers move their cattle to areas with shorter dry season in order to have better forage for cattle. In fact cattle were commonly moved from farm to farm (62% of total farms). However, the study results showed that herd sizes and stocking rates did not differ by season ($P<0.05$) (Table 5). It seems that this misconception occurred due to overlooking several important points: 1) needs of cattle removal in the beginning of the wet season for pasture recuperation, 2) farmers' limited capacity to move cattle, 3) existence of large farms with a small number of cattle, 4) farmers' annual perspectives for cattle feeding, 5) increased calving in the dry season, and 6) cattle removal for long-term cattle entrusting in the wet season.

It is apparent that natural forage for cattle is in shortage in the dry season. Farmers need to move their cattle from their farm, but in the beginning of the wet season. Farmers also need to remove their cattle for pasture recuperation in order to make pasture cover the grazing areas. It was observed that when the rain starts, some farmers remove weeds in their grazing areas and keep cattle out for a few months. These characteristics are probably more common in natural pasture species since natural pastures are rather susceptible to weed invasion. In the wet season, it is probably easier and cheaper for cattle owners to rent other pasture lands to keep the cattle when cattle can consume good forage.

In fact, it seemed difficult for many farmers (especially with small farms) to take their cattle to other farms far from their farms. It is probably because this practice is very costly or the number of cattle is not sufficient to make it worthwhile. If farmers have strong connections based on their family, it may be easier to control stocking rates by removing cattle in the dry season. However, the period of settlement in the study area was rather short (60% of farms less than 10 years, Yamamoto, 2004); therefore, it seems that farmers do not have reliable contacts to others where cattle can be kept in the dry season. Many of farmers in the area are probably not in a stabilization stage but in a survival stage.

Forty one percent of the farms did not move their cattle probably because they have large farms compared with the number of cattle. When cattle are grazed on other farms, cattle owners may need to watch the conditions of their cattle because cattle may consume low quality feed. Supervision are required in this culture when hiring others (Kaimowitz, 1996) and some farmers showed negative feelings about keeping cattle in other farms due to bad experiences. Moreover, farmers may plan dry season fodder from the wet season, thus some farmers may remove their cattle in the wet season for a long-term basis in order to keep their own pasture in the dry season.

For farmers who have excessive cattle with respect to the forage supply, one option is to sell animals. However, it seemed that they want to increase the number of cattle and are rather reluctant to sell the cattle (Yamamoto, 2004). Cattle are clearly an important investment for them; therefore, it seems that they do not sell their cattle even if they have to pay other farmers for caretaking.

In order to reduce the number of cattle for a long period of time, farmers were entrusting their cattle (their young heifers) to other farmers until they calved and their calves had been weaned (long term caretaking contracts). In general, cows returned to the owners with calves after weaning [usually half of the calves born, therefore this system is called "sistema media" (half system)]. The caretakers of the cattle take the milk produced and half of the calves born on the farm. It seemed that the system was less costly than using credits, which was not financially viable due to high interest rates (Holmann, 1999). The system has relatively low risks provided that there are reliable relationships between the two farmers. In the system, they share the risks because the cattle keepers take greater care of the cattle since they receive direct benefit from them. It was observed that cattle were moved on a long-term basis in the wet season when forage was abundant. Further studies are recommended on cattle movement and the role of local cattle sharing systems compared with credit systems available from local banks.

Finally, it is important to point out that sample farms were selected by accessibility, located near the road, where milk collectors can come most of the year. In fact, there were many farms inaccessible to milk collectors and these farms were probably available to keep cattle. On the contrary, the farmers may bring cows to the farms (possibly from others' farms) when they start producing milk to sell milk, resulting in maintaining large number of milking cows in the dry season when calving rates were rather high ($p<0.01$).

5.6. Supplementary Feeding and Farm Types

Size of grazing areas, stocking rates and the percentages of degraded/cultivated pastures did not differ significantly by classification based on supplementary feeding ($P<0.05$, Table 10). However, farms that used supplementary feeds, such as supplementary forages and molasses, tended to have higher stocking rates, more degraded pastures and less cultivated pastures in the grazing areas. It may indicate that some farmers with higher stocking rates prefer to invest in supplementary feeding than pasture improvement in the grazing areas.

CONCLUSION

The study results show that tree cover represented a large proportion of the land area on the dual-purpose cattle farms that were surveyed. Grazing lands with trees, so called silvopastoral areas account for 73% of the grazing areas. Cattle in the sample farms were frequently moved between local farms, but stocking rates did not differ significantly by season, suggesting that seasonal cattle movement towards more humid area is limited. The maintained stocking rates as well as significantly higher occurrences of calving causes serious fodder shortage in the dry season, resulting in lower saleable milk production per cow and per hectare as well as higher adult mortality rates. Smaller farms have higher stocking rates and are overgrazed, while the farms closely managed by owners started introducing cut and carry systems maintaining existing pasture resources. Farms with higher financial resource shift their production types to steer rearing which requires less labour inputs.

The study concluded that shortage of dry season fodder, lack of calving control and reluctance to increase payment for labour are the major obstacles for environmentally sound intensification of dual-purpose cattle farms. Further research is recommended on improvement of calving control, fodder availability in the dry season, feasibility of pasture improvement, cattle movement and sharing system, and relations between farm manager and paid labour.

ACKNOWLEDGEMENTS

The land use data and the field surveys of the study were benefited by the project, "Regional integrated silvopastoral approaches to ecosystem management project", undertaken by CATIE and NITLAPAN under the scheme of Global Environment Facility (GEF) of the World Bank. The authors wish to thank for providing the opportunity to undertake this study.

REFERENCES

Corporación Ganadería. (2000). Análisis de Censo Ganadero 2000. http://www.corfoga.org/pdf/proyecto/censo2000.pdf.

FAO (Food and Agricultural Organization). (2003). FAOSTAT.

Holmann, F. (1989). *Economic evaluation of dairy and dual-purpose cattle production systems in Venezuela*. Ph.D. Dissertation. Cornell University.

Holmann, F. (1999). Ex-ante analysis of new forage alternatives for farms with dual purpose cattle production in Peru, Costa Rica and Nicaragua. Livestock Research for Rural Development 11. http://www.cipav.org.co/lrrd.

Humphreys, L.R. (1987). *Tropical pastures and fodder crops*. Intermediate Tropical Agriculture Series. Longman Science and Technical. 414pp.

INTA.1998. *Zonificación agro socioeconómica agencia Matiguas*. Managua.

Kaimowitz, D. (1996). *Livestock and Deforestation: Central America in the 1980s and 1990s: A Policy Perspective*. CIFOR Special Publication.

Levard, L, Lopez, YM, Navarro, I. (2001). *Municipio de Matiguas: Potencialidades y limitantes del desarrollo agropecuario*. UCA/FIDA/NITLAPAN. Managua.

Llamozas, JA., Vaccaro, L. (2002). Correlations of part and total lactations and the prediction of lactation milk yield in Venezuelan dual purpose cows hand milked with calf at foot.

Livestock Research for rural development 14. http://www.cipav.org.co/lrrd/lrrd14/lrrd14.htm.

Maldidier, C, Marchetti, R. (1996). *El Campesino-Finquero y el potencial económico del campesinado nicaragüense*. NITLAPAN. Managua.

MINITAB. (2000). *MINITAB User's Guide 1 and 2.*

Minson, DJ. (1981). Nutritional differences between tropical and temperate pastures. In. Morley, FHW. (Eds). *Grazing animals. Disciplinary approach. World Animal Science. B1*, 143-157.

Murgueitio, E, Ibrahim, M, Ramirez, E, Zapata, A, Eduardo, C, Casasola, F. (2003). *Usos de la tierra en fincas ganaderas: Guía para el pago de servicios ambientales en el proyecto Enfoques Silvopastoriles Integrados para el Manejo de Ecosistemas*. CIPAV. Kali.

Myers, N. (1981). The Hamburger Connection: How Central America's Forest Became North America's Hamburgers. *AMBIO. 10*, 3-8.

Nicholson, CF, Blake, RW, Lee, DR. (1995). Livestock, deforestation, and policy making :intensification of cattle production systems in Central America revisited. *Journal of Dairy Science. 78*, 719-734.

NITLAPAN. (1995). *Diagnostico de la Producción Agropecuaria en el Interior del País. Análisis de Encuesta Rural.* Proyecto de Tecnología Agraria y Ordenamiento de la Propiedad Agraria. Managua.

Peters, M, Franco, LH, Schmidt, A, Hincapié, B. (2003). *Especies Forrajeras Multipropósito: Opciones para productores de Centroamérica*. CIAT. Kali.

Plasse, D. (1992). Presente y futuro de la producción bovina en Venezuela. In: González-Stagnaro, C. (Eds.), *Ganadería mestiza de doble propósito*. Universidad del Zulia. 1-23.

Parsons, J. (1976). Forest to Pasture: Development or Destruction? *Revista de Biología Tropical. 24*, 121-138.

Ramirez, A.C. (2002). *Ganadería de leche: Enfoque empresarial.* Editorial Universidad Estatal a Distancia. San Jose.

Ruiz, A. (1994). *Tenencia y Uso de la Tierra en Matiguas-Matagalpa, Nicaragua*. FAO Land Tenure Working Paper.

Sere, C, De Vaccaro, L. (1985). Milk production from dual-purpose systems in tropical Latin America. In: Smith, AJ. (Eds.), *Milk production in Developing Countries*. Centre for Tropical Veterinary Medicine. University of Edinburgh. 459-75.

Souza de Abreu, M.H. (2002). *Contribution of trees to the control of heat stress in dairy cows and the financial viability of livestock farms in humid tropics*. Ph.D. Dissertation. CATIE. Turrialba.

Stobbs, T.H, Thompson, PAC. 1975. Milk production from tropical pastures. *World Animal Review. 13*, 27-31.

Vaccaro. L, Vacarro, R, Verde, O, Alvarez, R, Mejias, H, Rios, L, Romero, E. (1992). Características productivas en la evaluación de explotaciones y vacas en sistemas de doble propósito. Turrialba. 42, 14-22.

Wilkins, JV, Pereira, G, Ali, A, Ayola, S. (1979). Milk production in the tropical lowlands of Bolivia. *World Animal Review. 32*, 25-32.

World Bank. 2004. Nicaragua at a glance.

Yamamoto, W. (2004). *Effects of silvopastoral areas on dual-purpose cattle production at the semi-humid old agricultural frontier in central Nicaragua*. Ph.D. Dissertation. CATIE/University of Wales, Bangor.

Yamamoto, W, ApDewi, I, Ibrahim, M. (2007). Effects of silvopastoral areas on dual-purpose cattle production at the semi-humid old agricultural frontier in central Nicaragua. *Agricultural systems 94*, 368-375.

In: Agricultural Systems: Economics, Technology and Diversity ISBN 978-1-60692-025-1
Editor: Oliver W. Castalonge

Chapter 9

GROWING MORE RICE WITH LESS WATER IN ASIA: IDENTIFYING AND EXPLORING OPPORTUNITIES THROUGH SYSTEM OF RICE INTENSIFICATION

Abha Mishra and V. M. Salokhe [*]
Agricultural Systems and Engineering
Asian Institute of Technology
Bangkok, Thailand

ABSTRACT

Due to ever-increasing population, water scarcity, and pressing issues of environmental sustainability, Asian rice farmers are under considerable pressure for sustainable increase in rice production by using less water. It is widely believed that an increase in the water use efficiency through integrated crop management holds promise of increased yields and water productivity. The so-called System of Rice Intensification (SRI) is attracting favorable attention of farmers and governments in Asia and elsewhere. It is assumed that a healthier and larger root system can be induced by using "SRI principles" in a water-limiting environment giving positive impacts on grain yield. The cultural practices that characterize SRI includes rapid and shallow transplanting of younger seedlings, at wider spacing and maintaining alternate wet and dry condition or preferably just moist conditions during the vegetative stage. This chapter reviews the biological mechanisms of water-saving agriculture and its relation to SRI cultural practices. It presents some research findings on the rice plant's adaptive trait which could be utilized to manage crops under limited water application. In addition, the details of on-farm studies carried out in some of the rice growing countries of Southeast Asia using a participatory action research approach are included in this chapter, which asserts the need for integrating science, people and policy makers for better and sustainable water management in Asia and elsewhere.

[*] E-mail: salokhe@ait.ac.th

Key words: agricultural water management, System of Rice Intensification (SRI), root growth, Participatory Action Research (PAR), Farmers' Field School (FFS).

INTRODUCTION

Asian farmers and rice growers in particular, are under considerable pressure to increase rice production sustainably by using less water. Major drivers are the growing population, water scarcity, reduced investment in irrigation infrastructure and pressing issues of environmental sustainability. While the water shortage in this region is severe, water use efficiency of rice is very low and is further lowered at the small farms which are supposed to contribute 75% of the additional food needed over the next decade. It is widely believed that an increase in the water use efficiency through integrated crop management holds promise to increase yields, improve water productivity leading to narrow the yield-gap experienced in these farmers' field in most of the developing countries of Asia. One, in particular, the so-called System of Rice Intensification (SRI) is attracting favourable attention of farmers and governments in Asia and elsewhere to mitigate theses challenges.

SRI, which is mostly empirical method of growing rice, is defined as a set of basic management practices (Stoop et al., 2002; http://ciifad.cornell.edu/sri) that includes:

- Transplanting very young seedlings (10-15 days old)
- Rapid and shallow transplanting of one or two seedlings with wider spacing (30 x 30 cm or 40 x 40 cm)
- Practicing alternate wetting and drying (AWD) during vegetative phases (or keeping the soil moist but not continuously saturated)
- Applying organic manure as much as possible

The SRI concepts, developed by Fr. Laulaniè in the 1980's in Madagascar, was intended to enable resource-limited farmers to obtain higher yields with less water and without primarily relying on external inputs for yield improvement. It sought to achieve this by altering the ways that rice plants, soil, water and nutrients are managed (Stoop et al., 2002). The cultural practices that characterize SRI such as rapid and shallow transplanting of younger seedlings, wider spacing and reduced irrigation water usage are all amenable to farmer experimentation and adaptation to suit local conditions. Thus, SRI encourages farmer participation in devising practical ways of growing a healthy crop in a sustainable manner.

In SRI, there are synergistic effects among the recommended practices, giving more increase in yield in water limiting environment when all are used together rather than used singly and separately. The synergistic effect of components of SRI practice rekindles some latent issues that are still not given adequate attention. One such critical issue is the contribution of root vigor towards yield (Mishra et al., 2006) especially in water limiting environment. It is proposed that a healthier, larger and prolonged active root system can be induced by using the SRI method in a water-limiting environment (Satyanarayana et al., 2007; Mishra et al., 2006), which is assumed to boost grain yield per plant. However, there is limited formal research activity investigating SRI and thus many have challenged the yield gains reported (McDonald et al., 2006). Given that agronomic practices are radically different from those promoted during the Green Revolution, the SRI provides unique opportunities for farmers, researchers, and extension workers to engage and be actively involved in exploring these newer agronomic practices. In particular, it is interesting to explore how these new agronomic practices can change the rice plant's phenotype and yield performance and how this can be achieved with reduced irrigation inputs so that farmers can obtain 'more crop per drop'. Certainly, this requires on-station experiments and on-farm studies to clarify various bio-technical issues, as well as to develop location specific technology for better adaptability of SRI practices.

This chapter reviews some of the biological mechanism of water-saving agriculture in relation to SRI cultural practices. The results of on-station research on some of the management practices of SRI, which have got major attention in Asia, is also discussed. In addition, the chapeter presents the outcome of on-farm studiescarried out using participatory action research in Cambodia and Thailand with aim to find synergies among various agronomic management practices in a given socio-economic conditions and to build capacity of the participants to manage a healthy rice crop in a water limiting environment.

Biological Mechanism of Water-Saving Agriculture and SRI Cultural Practices

A water-saving agricultural system refers to integrated farming practices that are able to sufficiently use rainfall and irrigation facilities for improved water use efficiency (Shan, 2002). The scientific measures in a water-saving agricultural system include spatial and temporal adjustment of water resources, effective use of natural rainfall, rational use of irrigation water and increased plant *water use efficiency* (WUE). In agricultural practices, several factors need to be take into account, namely, (i) the quantity, quality, spatial and temporal distribution of water resources, (ii) the establishment of cultivation practices aimed at reducing water consumption as a result of reshaping the existing farming structure and cropping systems in line with the current distribution pattern of water resources, (iii) sufficient manpower and equipment for the research, development, production, supply and maintenance of water-saving materials, spare parts, instruments and facilitates, (iv) relevant laws and statutes concerning water management to be enacted, formulated and perfected, and (v) a special campaign to enhance the public's water-saving awareness (Deng et al., 2003).

Among all, the establishment of newer cultivation practices, aimed at reducing water consumption as a result of reshaping the existing farming practices with better on-farm management can make significant contribution to 'grow more rice with less water' in Asia. A good understanding of factors limiting and/or regulating yield may provide researchers opportunity to identify and select physiological and breeding traits that increase plant WUE and drought tolerance under water-scarce conditions, where as better understanding and knowledge of soil-plant water relationship can help farmers to manage their rice crop with optimal water use. The details here basically deal with the physiological regulation of biological water-saving which might be possible to achieve using SRI cultural practices. The details of efforts to facilitate farmers for better understanding of soil-plant-water relationship through participatory action research conducted in Cambodia and Thailand for better on-farm water management are discussed too.

Regulating Photosynthesis by Increased Root Activity

Under field conditions there is a parabolic relationship between photosynthesis and transpiration (Wang and Liu, 2003), such that transpiration increases as photosynthesis increases but continues to increase when photosynthesis reaches a maximum so that ultimately transpiration efficiency decreases. At these levels, transpiration can be controlled without affecting the rate of photosynthesis if root activity is kept high (root activity is ά-naphthylamine-oxidizing activity of roots which is correlated with higher chlorophyll content of the leaves, slower leaf senescence, erect leaves and single leaf photosynthetic ability). Jiang et al. (1985) suggested that if root activity is kept high, photosynthesis doesn't decline much in the afternoon. By this way transpiration efficiency might be increased to enhance WUE. Based on that, a '*broken irrigation*' was introduced in the past to control tillering and to maximize efficiency at the harvest by regulating photosynthesis and by changing the sink-source relationship. In this method, after flooding the field, water intake was cut so as to let the water level go down gradually until the soil surface appeared. The field surface was exposed to air for three to five days, and then flooded again. A similar water management practice was recommended under SRI which suggests that these manipulations might have influence on root activity to regulate photosynthesis (Mishra et al., 2006). An experiment was conducted on rice in semi-field condition at Asian Institute of Technology (AIT), Thailand in 2006 on varying water regimes. Root activity and chlorophyll content of upper and lower leaf were studied under three water regimes:

i) Intermittent flooding (IFI) – Pots were maintained with IFI water regimes. The water depth was maintained at 5 cm everyday for 12 days, then drained for three days and again reflooded to the same depth of ponded water. Water was drained from the bottom of the pot by drainage hole and collected in container and was stored in a refrigerator during pot drying and returned to the pots when reflooding was done. Three, three-days-drying period were provided at 19, 34 and 50 DAT followed by flooded water treatment (5 cm water depth continuously) until maturity.
ii) Intermittent flooding (IFII) – In other pots, similar procedure like IFI was followed five times at 19, 34, 50, 66 and 82 DAT followed by flooded water until maturity.
iii) Continuous flooded (CF) – 5 cm depth of ponded water was maintained until maturity.

The experiment revealed that there was positive correlation between chlorophyll content of lower leaves and root activity in all water regimes (Figure 1). Significant and positive correlation was also found between chlorophyll content of flag leaves and duration of grain fillings (Figure 2). It was observed that senescence of lower leaf and flag leaf was delayed under intermittent irrigation applied during vegetative stage compared to continuous flooded condition and continuous intermittent irrigation (Figure 3). This delay was associated with higher root activity, and higher biomass production along with higher grain weight (Table 1). Therefore, it seems that intermittent irrigation recommended in SRI might decrease irrigation water requirement and would contribute towards higher yield when soil nutrients are not the limiting factor.

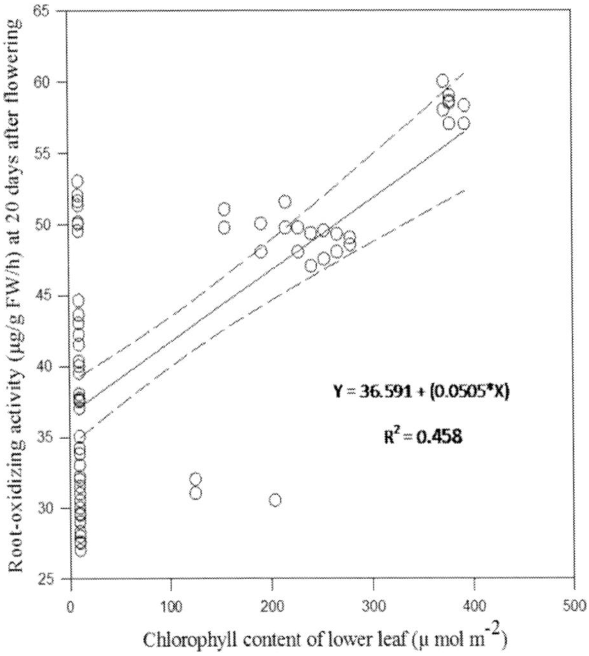

Figure 1. Linear regression slope between chlorophyll content of lower leaf (x) and root oxidizing activity at 20 days after flowering (y). Dotted lines show the 95% confidence level.

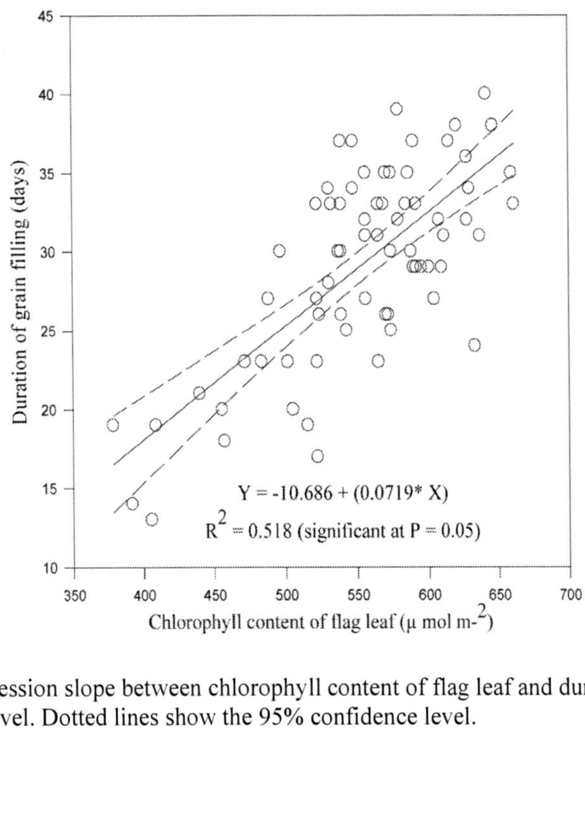

Figure 2. Linear regression slope between chlorophyll content of flag leaf and duration of grain filling at 95% confidence level. Dotted lines show the 95% confidence level.

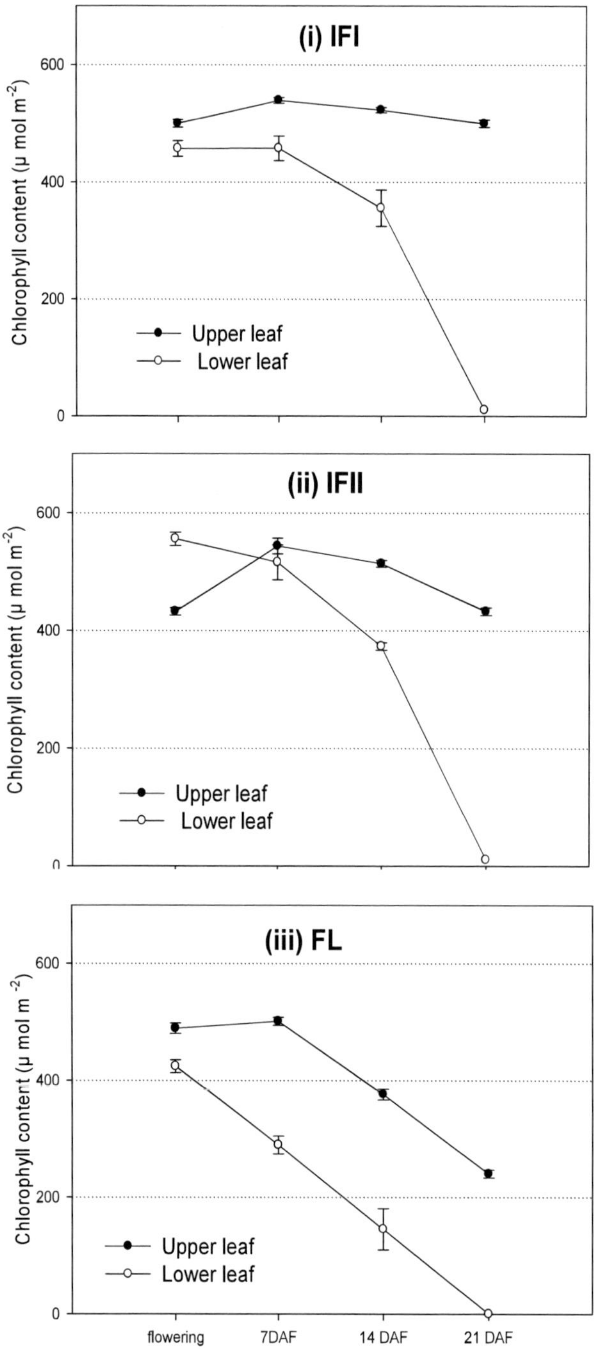

Figure 3. Changes in chlorophyll content of upper (flag leaf) and lower (3rd leaf) leaf of rice plant during reproductive phase maintained under three water regimes: IFI (intermittent drying for three times), IFII (intermittent drying for five times) and NF (nonflooded). Error bars show *s.e.* (n=2). For each replicate, three leaves were used for measurement. DAF denotes days after flowering.

Table 1. Changes in root activity, plant biomass, and grain weight under varying waterregimes in pot study conducted at AIT. (IFI – intermittent drying for three times; IFII – intermittent drying for five times; and CF – continuous flooding)

Treatment	Root oxidizing activity (μg/g FW/h) at flowering	Root oxidizing activity (μg/g FW/h at 20 days after flowering	Total biomass (g/plant)	Dry grain weight per pot (g)
IFI	63.40 ± 0.74	42.28 ± 0.57	341.71 ± 10.22	165.74 ± 4.07
IFII	53.10 ± 0.35	33.10 ± 0.33	215.73 ± 7.67	116.46 ± 2.19
CF	62.34 ± 0.56	30.12 ± 0.22	248.14 ± 9.75	101.75 ± 7.28

Means and *s.e.* at DF = 4, P< 0.05.

Regulating Plant-Water Relationship by Modifying Root Morphology

A deep root system is synchronous with more water uptake from the soil and better performance under drought. A deep and healthy root system is not only correlated with better water uptake but also influence yield physiology by regulating cytokinin production.

Cytokinin is regarded as the most important senescence-retarding plant hormone (Faiss et al., 1997). It is greatly regulated by environmental conditions of rhizosphere such as nitrogen availability, soil moisture condition, root mass, root length density etc. Among all, root length density and root mass are important variables for characterizing temporal trends in the water relations of rice especially when the water supply is scarce. The amount of water available to plant depends on the relative root length density and the ability of roots to absorb water from the soil.

Experiments conducted by Matsuki and Katsutani (1940) have shown larger root weight in lower soil layers under limiting moisture conditions. Baba (1977) also found that non-flooded soil condition in the early and mid stages of rice growth promoted the appearance of xeromorphic traits with increased root mass, and was consistent with physiological adaptation to drought conditions without compromising the yield. This was possible due to larger root growth both at upper and lower soil depths, and osmotic adjustment at early growth stage that helped plant to maintain photosynthesis with increased WUE. The simplest way to increase rooting depth and root distribution of crops is to increase the duration of the vegetative period. This may be achieved by sowing earlier or delaying flowering. SRI practice recommends transplanting of younger (12-15 days old) seedlings with wider spacing. This helps plants to have prolonged vegetative periods along with better canopy growth and enhanced canopy photosynthesis by avoiding shading effects. Intermittent irrigation or preferably 'just moist' soil condition at vegetative stage further helps plants to grow more roots with higher root length density at deeper soil layer. Yang et al. (2004) showed that intermittent irrigation increased the root length density, active absorption area, root oxidation ability, and nutrient uptake in rice.

Taking it to further, an experiment was conducted in 2006 at AIT in nursery seedbed and after transplanting to understand the roots and shoot characteristics of younger and older seedlings grown in wet and dry seedbed and their contribution to the production of tillers and dry matter. It was found that the root length density was affected by the age of seedlings, seedbed management and by water regimes at early growth stage. Flooded soil favored root length density at shallow soil depth mainly for older seedlings (Table 2). This could be due to preference of shoot growth over root growth in older seedlings and dominance of NH_4^+ in the soil solution which under reduced environment remains mostly in upper soil layer (Sah and Mikkelsen, 1983). In contrast, non-flooded soil improved root growth in the subsoil layer, but more in younger seedlings compared to older ones. The better uptake of N by younger seedlings grown in a dry seedbed was also seen. It was possible due to higher root length density and greater number of lateral roots that helped better acquisition of nutrients from the soil. Younger seedlings raised in a dry seedbed appeared to be the most suitable management practice to achieve higher tiller production along with higher biomass production even under non-flooded soil condition due to better root growth (Figure 4 and 5). This adaptive root trait could be exploited to manage rice crops under limited water application without compromising the grain yield (Mishra and Salokhe, 2008).

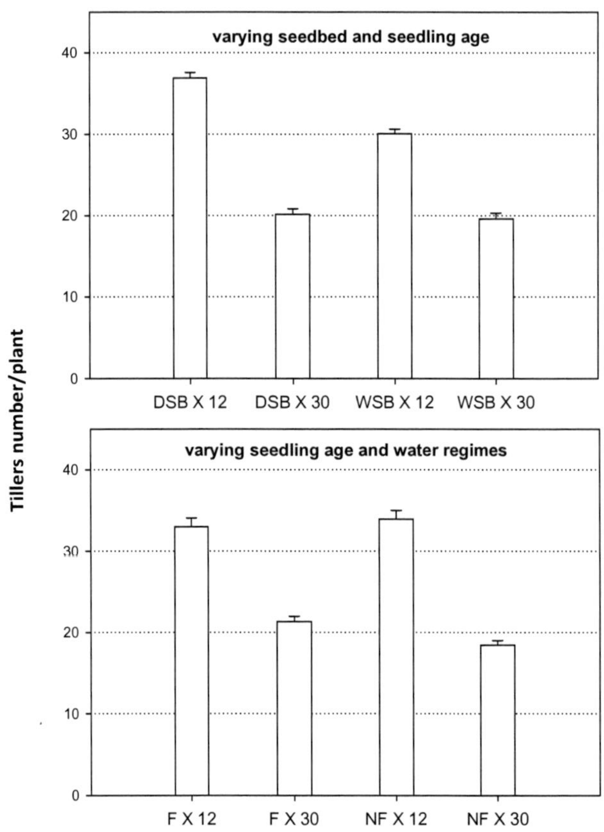

Figure 4. Plant's tiller development at 45 days after transplanting with interaction effect of different seedbed (DSB = dry seedbed and WSB = wet seedbed), seedling age (12 = 12 days old and 30 = 30 days old) and water regimes (F= flooded and NF = nonflooded). (N = 16). Error bars show *s.e.*

Table 2. Effects of seedbed management, seedling age, and water regimes on root characteristics and N uptake by rice plant at 45 days after transplanting in pot study conducted at AIT

Seedbed	Water regimes	RLD ($cm\ cm^{-3}$) upper soil layer		RLD ($cm\ cm^{-3}$) sub-soil layer		Total RLD ($cm\ cm^{-3}$)		N content in plant shoot (mg^{-pot})	
		12 days	30 days	12 days	30 days	12 days	30 days	12 days	30 days
DSB	F	5.78±*0.14*	4.55±*0.08*	1.97±*0.13*	1.30±*0.04*	7.75	5.84	321.75±*4.83*	253.88±*4.82*
DSB	NF	5.35±*0.18*	4.25±*0.16*	2.55±*0.14*	1.20±*0.08*	7.90	5.46	312.75±*7.68*	181.88±*2.45*
WSB	F	5.21±*0.13*	4.26±*0.17*	1.77±*0.12*	1.66±*0.11*	6.98	5.92	274.50±*9.04*	205.50±*2.53*
WSB	NF	5.14±*0.12*	3.22±*0.20*	1.92±*0.12*	1.79±*0.07*	7.06	5.01	243.75±*5.00*	184.63±*3.09*

RLD = Root length density. Mn and N content in shoot were expressed on dry weight basis.

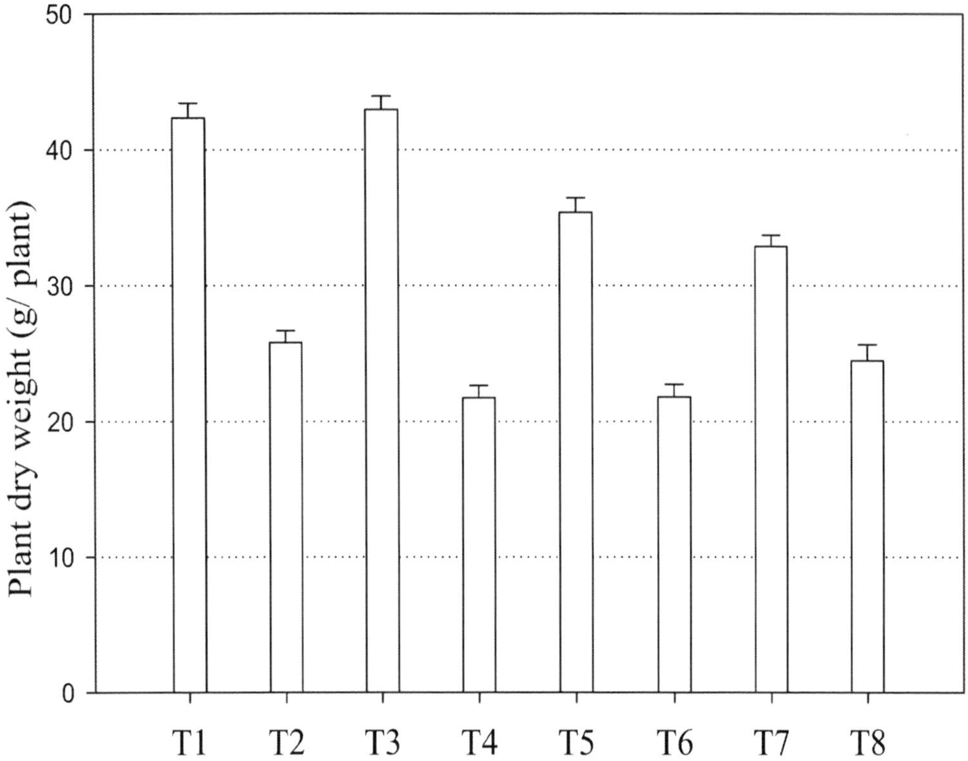

Figure 5. Plant biomass at 45 days after transplanting with effect of different seedbed, seedling age and water regimes. X axis shows the different treatments (T1= DSB+12+F; T2 = DSB+30+F; T3 =DSB+12+NF; T4 =DSB+30+NF; T5 = WSB+12+F; T6 = WSB+30+F; T7 = WSB+12+NF; T8 = WSB+30+NF). (N = 8) Error bars show *s.e.* (See figure 4 for notations used).

Based on the above findings it may be stated that a healthier, larger and prolonged active root system can be induced by using the SRI cultural practices in a water-limiting environment. However, these remain to be tested over different locations with different variety before making any general conclusions, due to limited formal research activity investigating SRI systematically.

Engaging Farmers, Researchers and Extension Workers through SRI for Better On-Farm Management for '*More Crop Per Drop*'

Above reviews and some initial research findings indeed provide some opportunity to manage rice crop by optimizing soil-plant-water relationship in a water limiting environment. However, given the agro-ecological and socio-economic diversity of rice production system, there is a need to develop knowledge-intensive and location-specific technology to realize the merit of SRI crop management principles.

In recent year, there has been an increase in the number of publications about SRI However, there is no disagreement that there still exists a significant gap between on-research-station yields and those obtained by most Asian farmers in their own paddy fields (Stoop and Kassam, 2005; Mc.Donald et al., 2006). This gap and current economic exigencies along with water scarcity provide incentives for farmers to explore the potential of novel options offered by SRI. It offers a set of management practices that farmers can evaluate, adapt and then adopt to meet their own local requirements. Moreover, it relies minimally on external inputs and maximally on farmers acquiring and using new knowledge and skill.

Yield increase at farmers' field under SRI management compared to conventional practice has been reported repeatedly (Koma, 2002, Anthofer, 2004). Even though more and more farmers are actively engaged in evaluating and adopting SRI practice, many knowledge gaps remain, and robust investigations might help to understand the SRI phenomena. In particular, such process of inquiry should focus on 'location-specific' management practices for optimizing soil-plant-water relationship for better yields with minimal water application. Given that agronomic practices are radically different from those promoted during the Green Revolution; the SRI provides unique opportunities for farmers, researchers, and extension workers to engage-in and be actively involved- in exploring these newer agronomic practices in general, and growing healthier root systems, in particular. Such collaborative investigation efforts require common and viable platforms at the grass-root level. Farmers' Field Schools (FFS), an approach developed and promoted by FAO to educate rice farmers on Integrated Pest Management (IPM) in Asia (Kenmore, 1991) offer suitable platforms for such investigative partnerships to explore SRI.

Keeping these concepts and facts in mind, a collaborative Participatory Action Research (PAR) initiative was set up in Prey Veng, Cambodia during 2005/6 under overall coordination of the National IPM Program of Ministry of Agriculture, Forestry and Fisheries with funding support from Regional IPM Program of Food and Agriculture Organization of the United Nations.

Two major factors – water and planting were identified by participants. Subsequently, a field trial was designed at Prey Veng province of Cambodia. The main factor was water regimes -- flooded (FL), alternate wet and dry (AWD), and just moist (JM), and the sub-factor was planting densities – six seedlings/hill with narrow spacing (6; 15 x15 cm), single seedling with narrow spacing (1; 15 x 15 cm) and single seedling with wider spacing (1; 30 x 30 cm). In 'FL' treatment, 5 cm water depth from the soil surface was maintained until panicle initiation stage and then 10 cm water depth was followed until 7 days before harvesting. In 'JM' moisture regimes, soils were kept continuously moist by irrigating plots with 1-1.5 cm water depth from the soil surface and re-applied when the top-soil surface started to dry up. In 'AWD' moisture regimes, intermittent irrigation was followed with initial 5 cm water depth from the soil surface and re-irrigated when the soil had developed fine cracks. In both 'JM' and 'AWD' water treatments; irrigation was stopped 20 and 25 days respectively before harvesting. –The single seedling with wider spacing under alternate wet and dry water regimes was treated as SRI where as six-seven seedlings/ hill with 15 x 15 cm spacing under flooded water regimes was treated as conventional practice.

After transplanting, observations on above-ground plant parts were made during regular Agro-Ecosystem Analysis (AESA) sessions that included yield attributing characters (per hill and per m^2) such as plant height, number of tillers/hill, number of leaves/tiller, leaf color at vegetative and reproductive stage, and panicle length, number of productive tiller/m^2, number of grains/panicle, number of filled grains/panicle, thousand grain weight, and yield/plot at harvest time. In addition, observations were made on the water level in each plot, insect pest populations, and on weather conditions. To correlate above-ground plant parameters and their related findings to the below-ground plant parts – roots were observed at seedling, tillering, flowering and harvesting stage by developing root AESA.

Based on the agronomic and AESA results, the participants involved in this action research concluded that:

- Using less water under 'just moist' field condition more number of productive tillers could be produced (Figure 6).
- Less water (under JM and AWD) at later growth stage accelerate grain filling (Figure 7), so this gives further opportunity to produce higher yields with less water inputs.
- A single seedling/hill is better than 6-7 seedlings/hill in terms of getting higher grain yield with better economic returns (Fig 8). Generally, farmers use 35-45 kg seed/ha for transplanting whereas for using the single seedling method – 6-7 kg seed/ha would be sufficient.
- At vegetative stage, the rice plant grows equally better in just moist or alternate wet and dry water regimes compared to those grown under continuous flooded regimes; hence, water inputs could possibly be reduced.

Results clearly demonstrated that SRI effect on rice yield is not governed by single factors but there are many factors which act simultaneously and in a synergistic mode. Such system understanding can only be possible when farmers are able to understand these interaction effects. IPM farmers, who have gone through season-long Farmers Field Schools (FFS), have developed such ecosystem thinking and management skills through discovery-learning (van de Fliert and Braun, 2002). In this trial, the IPM principles were followed by using "Agro-ecosystem Analysis" (AESA) -- a useful training tool, that focuses on farmers making regular field observations and collecting data on the plant and its micro-environment. Data are then analyzed and summarized on a so-called Agro Ecosystem Analysis (AESA) poster followed by plenary group discussions with the aim to arrive at informed crop management decisions. In this SRI investigation, AESA helped farmers to discover the fundamental causal relationship between roots and shoots of the rice plant which varied under different planting densities and water regimes. In particular, these investigations assisted farmers to understand why ensuring the growth of healthy root systems has the potential of boosting yields in water limiting environment. This had truly been innovative and resulted into the development of adapted FFS training curricula and learning exercises on growing healthy root systems.

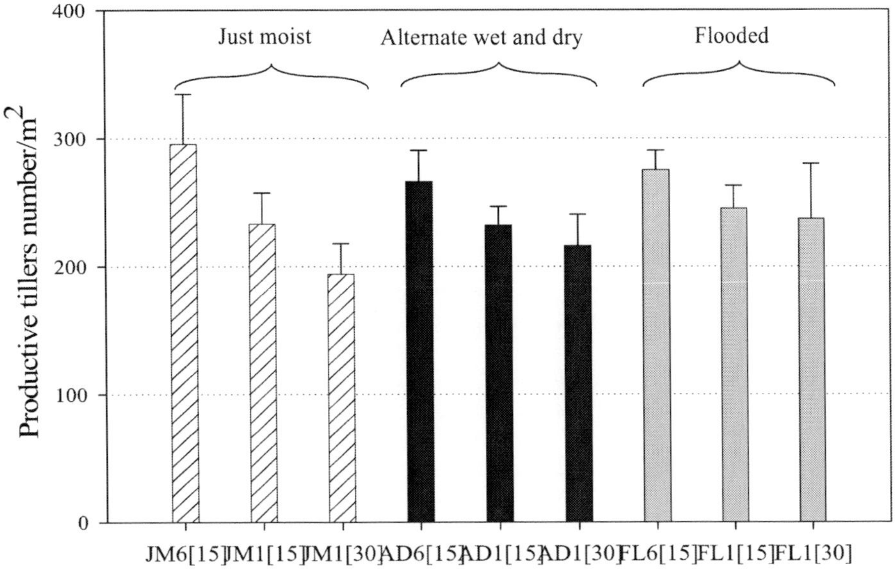

Figure 6. Highest number of productive tillers under JM water regimes in participatory action research (PAR) trial conducted in Cambodia. JM6[15] represents six seedlings with 15 x15 cm spacing under just moist (JM) water regimes; JM1[15] is single seedling with 15 x 15 cm spacing under JM; JM1[30] is single seedling with 30 x 30 cm spacing under JM water regimes. Similarly AD and FL are alternate wet and dry and flooded water regimes respectively. Error bars show *s.e.*

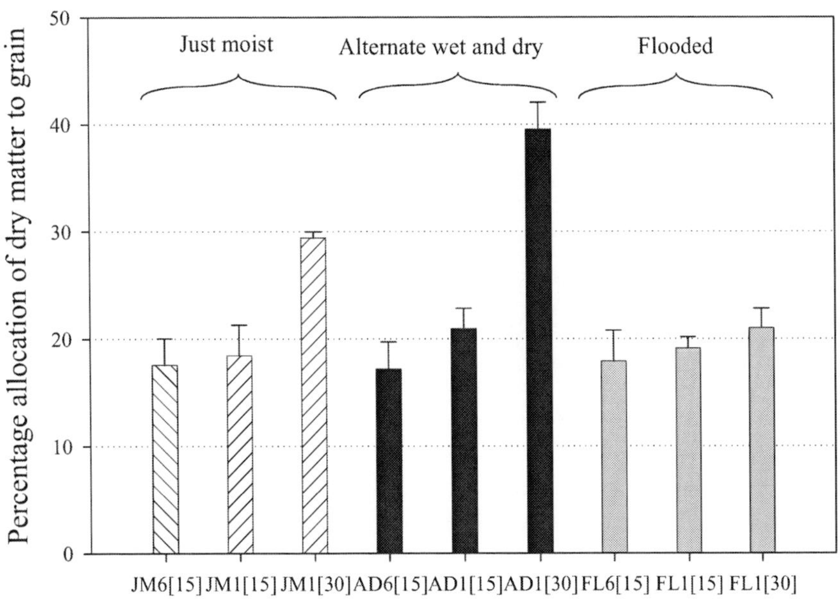

Figure 7. Higher percentage of dry matter allocation to the grain in single seedling with 30 x 30 cm spacing grown in nonflooded water regimes (JM and AWD) in PAR trial conducted in Cambodia. (See Figure 6 for details of symbol on X axis). Error bars show *s.e.*

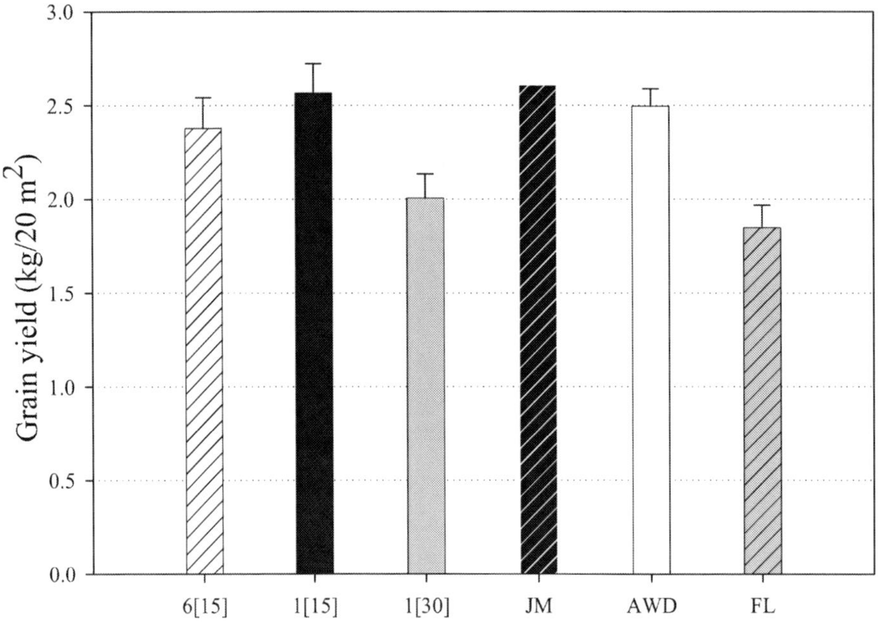

Figure 8. Grain yield at 14% moisture content in different water regimes and planting densities in PAR trial conducted in Cambodia. (See Fig. 6 for details of symbol on X axis). Error bars show *s.e*. Please note that FL (flooding) is simulated for conventional practice whereas, nonflooded (AWD or JM) is simulated for SRI method. There was no interaction between water regimes and planting densities for grain yield.

A similar collaborative enquiry into the water productivity issues of the transplanted rice under the ambience of SRI were carried out with a group of farmers, Non Government Organization (NGO) and Government Organization (GO) personnel in Ban Chaeng, District, at Samart, Roi-Et, Thailand using a participatory action research with a funding support from the Consultative Group of International Agricultural Research (CGIAR) through 'Challenge Program for Water and Food' grant with project Title "Increasing Water use efficiency in Rice using Green mulch under SRI management practice". 'Water use in rice' was discussed with the participating farmers and non-formal education trainees during weekly FFS conducted for 18 weeks. Two experiments were carried out during first season in Wet season 2006, and in experiment-1, where the two water regimes i.e., Just moist (JM) was compared with the farmers' practice (flooding), no significant difference in crop yields were noticed. The JM produced similar rice yield per unit area with less supplementary irrigation. In experiment 2, where SRI practice was followed in combination with green mulch, SRI and Mung Bean combination was proving to be the best among all other tested bean intercropping, thereby providing high foliage and ground cover as green mulch to the rice crop grown under SRI method. The results of experiments are shown in Figures 9-12. The results clearly showed that higher water use efficiency and rice yield obtained under SRI + green mulch (e.g. Mung bean) under just moist water regimes, and performed better than any existing farmer's practice. Also the results confirmed the benefit of younger seedling transplant (14 days) over older seedling (30 days) in another set of trial.

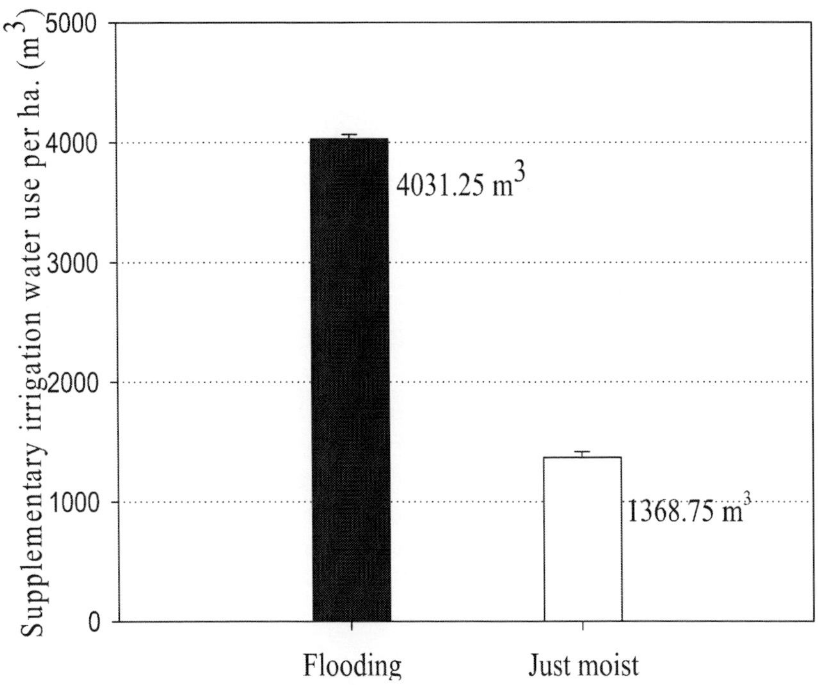

Figure 9. Total volume of supplementary irrigation water used under just moist and flooded rice cultivation system in PAR trial conducted in Thailand. Please note that this information is compiled to show the difference of water-use in two systems of rice; flooding (traditional system) and Just moist i.e., non flooded water regimes (SRI method). Error bars show *s.e.*

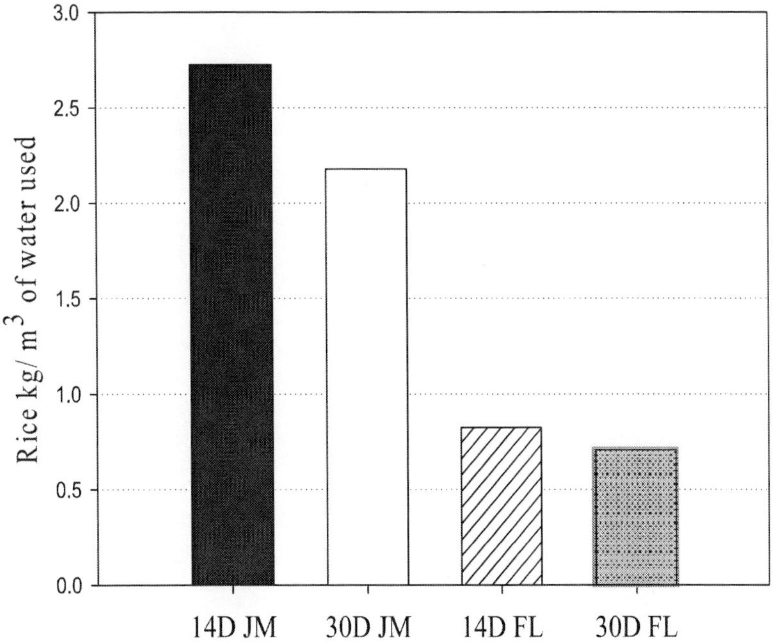

Figure 10. Water productivity for two tested conditions of water regimes in PAR trial conducted in Thailand. Please note that the 30 days old seedling and flooded water regimes (30D FL) are simulated as conventional farming practice used in the experiment for comparison purposes.

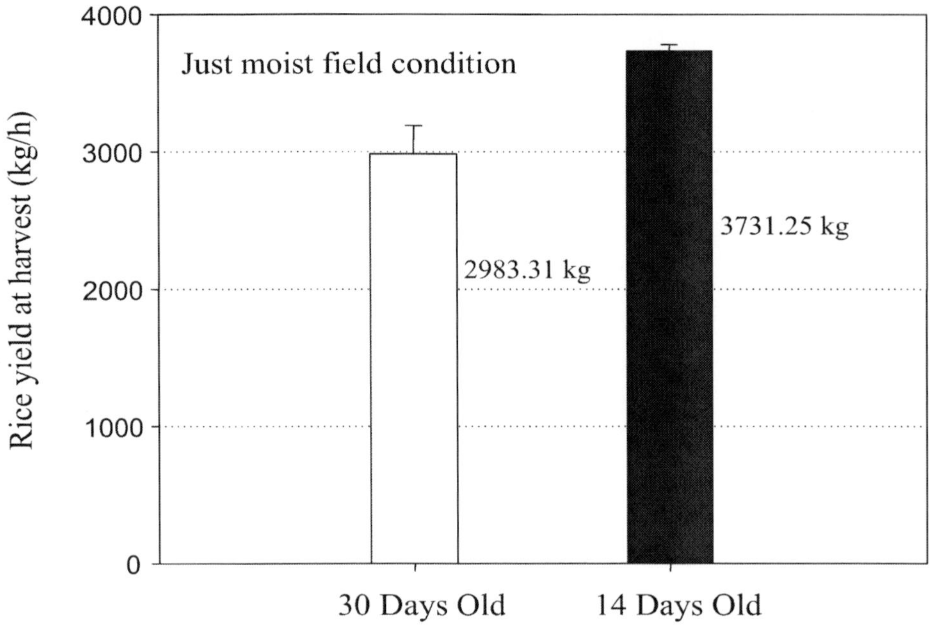

Figure 11. Rice yield under just moist (JM) condition in PAR trial, Thailand. 14 days old seedling performed better over 30 days old seedling under similar water and other management conditions. Error bar shows *s.e.*

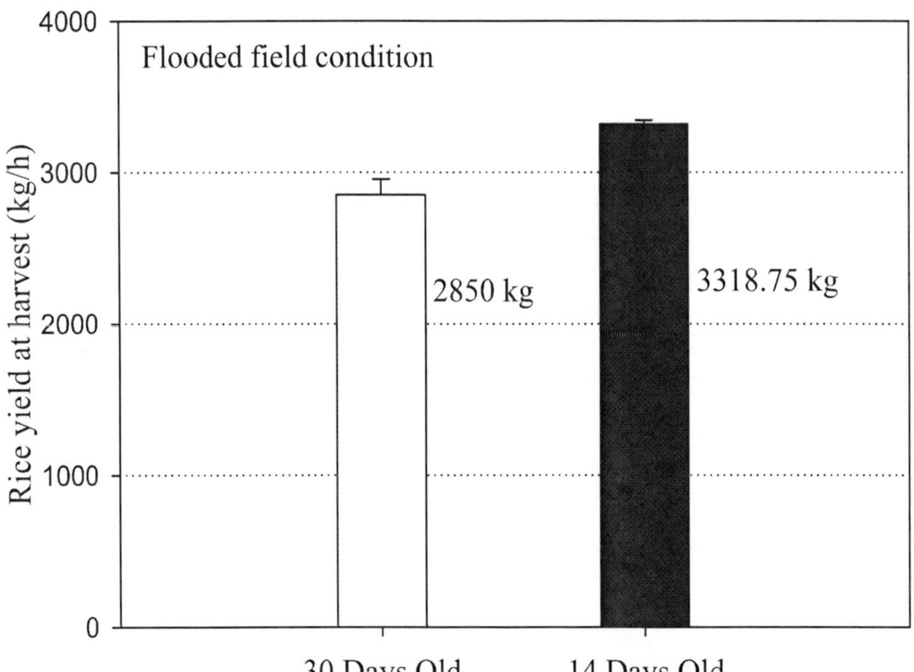

Figure 12. Rice yield under flooding condition in PAR trial, Thailand. 14 days old seedling performed better over 30 days old seedling under similar water and other management conditions. Error bars show *s.e.*

Yields obtained in either of the location in Cambodia and Thailand exceeded the average rice productivity in Prey Veng and Roi-Et province respectively. Indeed, it confirmed that SRI focuses on good crop husbandry and could be seen as a sound alternative management practice to minimize the yield gap experienced in most of the farmers's field in Asia by optimizing inputs use such as water and seed.

The major difference between SRI and the current 'input-driven' rice farming is that, SRI encourages manipulation of plant's aerial and soil environment by changing planting density, water and soil management rather relying on external physical inputs, thus encourages optimization of interaction effect of genetic endowment (G) and environment (E) often termed as 'G x E' effect, and results into healthy plants and healthy roots leading to higher yield. Such optimization requires skills for manipulating internal resources. This could be possible only when farmers are able to understand and appreciate the importance of soil-plant-water relationship for better production.

Hence, given the nature of SRI's focus on system thinking and farmer management skills, empowered IPM farmers appear ideally suited for exploring SRI and capturing potential benefits of such explorations. Given the widening gap between potential farm yield and actual farm yield in Asia (McDonald et al., 2006), such investigations and strengthening of farmers' management and inquiry skills are more urgently needed than ever.

Conclusion and Policy Implications

The need to grow more rice with less water in a more sustainable production environment is a big future challenge for Asian farmers and supporting R&D systems. SRI appears to provide a good opportunity to explore the contribution of "hidden-half" to yield through basic research by addressing soil-plant-water relationship. Even so, the practical constraints in managing favorable interaction of plant-soil-water environment should not be underestimated. However, these constraints could be minimized if SRI is used as a heuristic vehicle for linking rice research with participatory farmer empowerment programs. The SRI should be seen as a unique opportunity to integrate science with the society for sustainable development to address the 'rice demand with less water' in an integrated manner by focusing on improving farmers' production management skills and involving them in on-farm participatory research.

The lessons learned from the work reported in this chapter include:

- Based on the findings from on- station research trial, it can be assumed that it is possible to increase rice yield by enhancing the physiological efficiency of a rice plant by increasing root activity under limited water supply. Further research should be conducted to unlock the contribution of other roots traits which are strongly linked to the biomass production and grain yield.
- The better SRI results obtained by Cambodian or Thailand farmers suggest that SRI agronomic practices do make sense and deserve further on-farm investigation and promotion within agricultural training/extension programs that focus on developing farmers' production management skills.

- Informed management is needed to realize the synergistic effect of soil-plant-water relationship for better on-farm SRI adaptation. The FFS approach of farmer education appears most suitable for developing such farmer management skills.
- Finally, there is a need for further sensitization of researchers and scientists towards the potential benefits to be captured from more participatory and action oriented research approaches. SRI appears to be ideally suited for such research initiatives.

ACKNOWLEDGEMENTS

Research trials conducted at AIT were partly supported by a grant from the Asia Rice Foundation, USA, 2005. Funding and logistic support from FAO regional vegetable IPM Program for participatory action research on SRI and Soil Ecology in Cambodia is greatly acknowledged. We sincerely thank CGIAR for providing funds for participatory action research set up in Thailand.

REFERENCES

Anthofer, J. (2004). The Potential of the System of Rice Intensification (SRI) for Poverty Reduction in Cambodia. In "*International Agricultural Research for Development". Berlin*, October 5-7.

Baba, I. (1977). Effects of water stress on the physiology and the growth of paddy rice plants in relation to the generation of ethylene. *Japanese Journal of Crop Science.* 46 (1):171-172.

Deng, X. P., Shan, L., Zhang, S.Q., Kang, S. Z. (2003). Outlook on plant biological water-saving strategies. In: *Proceeding of KRIBB Conference on Environmental Biotechnology,* October 21–23. Daejeon, Korea.

Faiss, M., Zalubilova, J., Strnad, M., and Schmulling, T. (1997). Conditional transgenic expression of the IPT gene indicates a function for cytokinins in paracrine signaling in whole tobacco plants. *Plant Journal.* 12: 401- 415.

Jiang, C. Z., Hirasawa, T., and Ishihara, K. (1985). Eco-physiological characteristics of two rice cultivars, photosynthetic rate, water conductive resistance and interrelationship between above and underground parts. *Japanese Journal of Crop Science.* 57: 132-138.

Kenmore, P. E. (1991). 'How rice farmers clean up the environment, conserve biodiversity, raise more food, make higher profits in Indonesia's integrated pest management - A model for Asia' FAO Inter-country Program for the Development and Application of Integrated Pest Control in Rice in South and South-east Asia. (FAO: Metro Manila Philippines).

Koma, Y. S. (2002). Experiences with the system of rice intensification in Cambodia. In: Uphoff, N., Fernandes, C.M., Longping, Y., Jiming, P., Rafaralahy, S. and Rabenandrasana, J. (Ed.): *Assessment of the system of rice intensification* (SRI). Proceedings of an International Conference, Sanya, China, April 1-4, 2002, CIIFAD, Cornell, USA, 83-85.

Matsuki, G. and Katsutani, S. (1940). Some chemical investigations on the drought damage. 1. Soil moisture and plant growth. *Journal of Science and Soil Manure*, Japan. 14: 279-288.

McDonald, A. J., Hobbs, P. R., and Riha, S. J. (2006). Does the system of rice intensification outperform the conventional best management practices? A synopsis of the empirical record. *Field Crop Research.* 96 (1): 31-36.

Mishra, A and Salokhe, V. M. (2008). Seedling characteristics and the early growth of transplanted rice under different water regimes. *Experimental Agriculture* 44 (3): 1-19.

Mishra, A; Whitten, M; Ketelaar, J. W. and Salokhe,V. M. (2006). 'The system of rice intensification (SRI): a challenge for science, and an opportunity for farmer empowerment towards sustainable agriculture'. *International Journal of Agricultural Sustainability*. 4(3):193-212.

Sah, R. N. and Mikkelsen, D. S. (1983). Availability and utilization of fertilizer nitrogen by rice under alternate flooding. II: Effects on growth and nitrogen use efficiency. *Plant and Soil.,* 75: 227-234.

Satyanarayana, A., Thiyagarajan, T. N., Uphoff, N. (2007). Opportunities for water saving with higher yield from the system of rice intensification. *Irrigation Science.* 25: 99-115.

Shan, L. (2002). Development of tendency on dry land farming technologies. *Agricultural Science of China.* 1: 934–944.

Stoop, W. A., and Kassam, A. H. (2005) The SRI controversy: a response. *Field Crops Research.* 91: 357-360.

Stoop, W. A., Uphoff, N., and Kassam, A. H. (2002). A review of agricultural research issues raised by the system of rice intensification (SRI) from Madagascar: Opportunities for improving farming systems for resource-poor farmers. *Agricultural Systems.* 71: 249-274.

Van de Fliert, E. and Braun, A. R. 2002. Conceptualizing integrative, farmer participatory research for sustainable agriculture: From opportunities to impact. *Agriculture and Human Values* 19: 25–38.

Wang, H. X., Liu, C. M. (2003). Experimental study on crop photosynthesis, transpiration and high efficient water use. *Chinese Journal of Applied Ecology*, 14: 1632–1636.

Yang, C., Yang, L., Yang, Y., and Ouyang, Z. (2004). Rice root growth and nutrient uptake as influenced by organic manure in continuously and alternately flooded paddy soils. *Agricultural Water Management,* 70: 67-81.

Index

A

B

C

F

G

H

I

J

K

L

N

O

P

S

T

U

V

W

Y

Z